STATICS
FOR STUDENTS

EMERY BALINT

M.C.E.(Melb.), M.I.C.E.
Associate Professor of Building
University of New South Wales

Springer Science+Business Media, LLC

ISBN 978-1-4899-6179-2   ISBN 978-1-4899-6359-8 (eBook)
DOI 10.1007/978-1-4899-6359-8

©

Springer Science+Business Media New York 1967
Originally published by Butterworth & Co. (Publishers) Ltd in 1967.
Softcover reprint of the hardcover 1st edition 1967

Suggested U.D.C. number 531.2

Library of Congress Catalog Card Number 67–31094

NOTES ON STANDARD NOTATIONS

$A, B, C \ldots$	points or sections of a body
$F_A, F_B, F_C \ldots$	forces acting at A, B, $C \ldots$ respectively
x, y and z	co-ordinates of a point on the line of the force
F_{Ax}, F_{Ay}, F_{Az}	x, y and z components of the force acting at A
F_{Bx}, F_{By}, F_{Bz}	x, y and z components of the force acting at B
$R_x; R_y, R_z$	x, y, z components of the resultant force at the point of transposition
M_x, M_y, M_z	x, y, z components of the resultant moment at the point of transposition

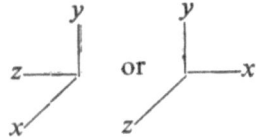 standard, right-hand screw co-ordinate systems

Figures in examples and in answers to problems were given to two or three digits, with slide rule accuracy.

Self-weight of a structure is *included* in the loads specified unless it is specially mentioned.

The homeless ferment of the universe
Finds equilibrium and repose
In the languid pulsing of a butterfly,
The sculptured limits of a pouting rose.
Beyond the laughter of unbridled girls
A descant sings, and sorrow is the air.
To the foundering heart awash with recent grief
There comes a point of balance with despair.
For balance is the buoyancy we live in,
To every pull an equal counterpoise.
The tension of opposing forces brings
A state of rest, a pause that earth employs
To harbour life. . . .
. . .

Robert Clark,
"Newton's third law of motion"

CONTENTS

		Page
PREFACE		xi

Chapter

1. BASIC CONCEPTS IN STATICS — 1

1.1.	The place of Statics	1
1.2.	Forces and moments	2
1.3.	External forces	6
1.4.	Internal forces	9
1.5.	Concentrated and distributed forces	16
1.6.	Pressure and stress	18
1.7.	Summary: Basic concepts	24
1.8.	Problems	24

2. MANIPULATION OF FORCES — 30

2.1.	Forces are vectors	30
2.2.	Two force vectors in a plane	32
2.3.	Two vectors in space	34
2.4.	Three force vectors in one plane	37
2.5.	Three vectors in space	43
2.6.	Four or more force vectors: the method of transposition	47
2.7.	Choice of co-ordinate system	58
2.8.	The Balancing Force is opposite to the Resultant	59
2.9.	Summary: Vector manipulation	62
2.10.	Problems	63

3. PRINCIPLES OF EQUILIBRIUM — 70

3.1.	Effect of supports on internal forces	70
3.2.	Types of support	72
3.3.	Supports in practice	77

3.4.	Special cases: plane structures and loads	78
3.5.	Plane supports in practice	84
3.6.	Summary: Principles of Equilibrium	85
3.7.	Problems	87

4. BODIES IN EQUILIBRIUM — 96

4.1.	Stability and statical determinacy of a rigid body	96
4.2.	Articulated bodies	101
4.3.	Inserted hinges	103
4.4.	Fully articulated structures	111
4.5.	Stability of plane trusses	112
4.6.	Stability of space trusses	118
4.7.	Articulated frames	122
4.8.	Summary: Stability and statical determinacy	124
4.9.	Problems	125

5. EQUILIBRIUM OF PARTS — 133

5.1.	Revision: Equilibrium of the entire body	133
5.2.	Internal forces: Definitions and conventions	134
5.3.	Concentrated loads	140
5.4.	Uniformly distributed loads	148
5.5.	Miscellaneous and mixed loadings	153
5.6.	Internal forces in frames	159
5.7.	Internal forces in trusses	166
5.8.	Equilibrium of pin method	167
5.9.	The method of sections	175
5.10.	Summary: Equilibrium of parts	182
5.11.	Problems	182

INDEX — 189

PREFACE

This is a textbook planned and written for the student with the aim of providing a clear explanation of equilibrium.

In the beginning, the book introduces the student to the concept of equilibrium as it exists in the mechanics of solids and fluids. The science of Mechanics is based on *forces* and we explain the origin and nature of forces. Statics deals with *systems of forces* which are in equilibrium and we must be able to manipulate these systems in order to establish equilibrium.

Once we have understood the basic principles of Statics, we can proceed to the solution of problems. For bodies in equilibrium, the total force system which is made up of acting loads and support reactions, must be in equilibrium. This consideration leads us to the finding of reaction components for solid bodies which support external loads.

Equilibrium is only possible for *stable* bodies, such bodies remain in equilibrium under the action of any load system of reasonable magnitude. On the other hand, a body may be *over-stable* if it has more than the required support restraints, or members, to make it *just stiff*. These *redundant* bodies may still be in equilibrium but their solution requires the rules of Elasticity, whereas the book restricts its treatment to cases of Statics. We differentiate, then, between unstable and stable bodies and also between statically determinate and redundant bodies. We will mainly deal with *stable* bodies which are built and supported in a *statically determinate* manner.

Once these external conditions are known, we are ready to *analyse* the body under the action of loads. This involves the finding of *internal forces* throughout the body; indeed without knowing these internal force components, the designer would not be informed on the behaviour of his structure or mechanism under load. This, then, is one of our final aims: to describe the internal forces in all sections of the loaded body.

Statics alone will not provide all the information for design: properties and strength of materials will also be of vital importance. But the understanding of the principles of equilibrium is of great importance and it is essential before the solution of complex structures can be attempted.

These principles apply both in *space* and *plane*. In this work, the student is encouraged to think in three dimensions right from the start and, wherever possible, the planar structure is regarded as just a special case of the general space body.

To make this approach easier, a systematic method is introduced, called the *method of transposition*. This can be used in all problems of Statics from finding reactions to applying the method of sections in space trusses. We choose a co-ordinate system and carry out all our calculations within this framework. All the work goes into a standard tabulation which makes these calculations almost as foolproof as using a slide rule for multiplication. There we hardly ever think of the way in which it works, that it converts a product into the antilog of a sum of two logs. Here, too, we may automatically fill in rows and columns in a table with hardly a thought for the principle of transposition, the basis for it all. Since the tabulation can be programmed for computer use, the operation could become quite mechanical.

All this brings us back to the need for *understanding*. Students should become familiar with the essential principles before they are confronted with complex problems and routine solutions. The worked-out examples and problems of the book all serve this main purpose: to throw light on the basic matter rather than to coach the students to pass tests or examinations.

The book has been written for the student in engineering, architecture or building. It is taken for granted that university students would have studied some of the introductory matter in preparatory subjects such as Physics. Its inclusion was found to be essential in order to build up the student's understanding of the principles of equilibrium and their application to three-dimensional structures.

The author wishes to thank Samuel Aroni, Associate Professor of Engineering, San Francisco State College, for assistance and criticism; John G. Balint, M.Eng.Sc.(Melb.), for providing some of the problems of Chapters 1 and 2, and his wife Eva, for typing and checking the manuscript.

EMERY BALINT

CHAPTER 1

BASIC CONCEPTS IN STATICS

1.1. THE PLACE OF STATICS

The science of *Mechanics* is of great importance in the study of Engineering and it is often used in the solution of technical problems. In a broad definition: Mechanics describes the behaviour of matter at rest or in motion, under the effect of acting forces.

Experiments and observations are easily made on phenomena in Mechanics and this is why Mechanics was one of the earliest of the physical sciences and its principles were the first to be formulated. In the third century B.C., Archimedes defined the laws of buoyancy and the principles of the lever. But it was not until the sixteenth century, 1,900 years later, that the principles of equilibrium were formulated by Stevinus. Rapid advances in the Science of Mechanics now followed and Galileo, Newton, Euler and others contributed to its development.

The science of Mechanics includes *Statics* which deals with the action of forces on bodies in equilibrium, and our textbook is restricted to this subject. Other branches of Mechanics describe the behaviour of bodies which are not in equilibrium: *Kinematics* deals with the motion of a body without considering its mass or acting forces and *Dynamics* includes consideration of both the mass and the acting forces.

A further division of Mechanics can be made into 'Mechanics of Solids' and 'Fluid Mechanics', with 'Soil Mechanics' or 'Grain Mechanics' occupying an intermediate position. A convenient family tree of Mechanics is shown in *Figure 1.1*.

Classical Statics deals only with the conditions which

1

determine whether a body is in equilibrium and it is *not concerned with the effect* produced by the forces which act on the body (such effects are deformation or strain). In consequence, in Statics no reference needs to be made to material properties but we may deal with *rigid bodies*.

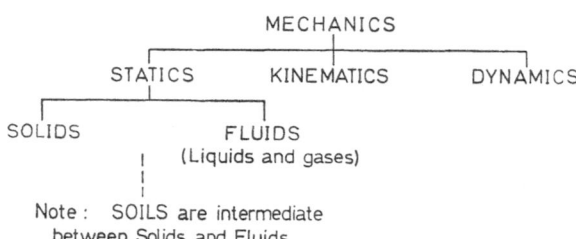

Figure 1.1. Main branches of Mechanics

1.2. FORCES AND MOMENTS

Force is the action of one body on another. According to Newton's Second Law, the force *F*, acting on a body, will accelerate it and the acceleration, *a*, will be proportional to the force:

$$F = m \times a$$

m, the constant of proportionality, is termed the *mass* of the body (*see Figure 1.2*).

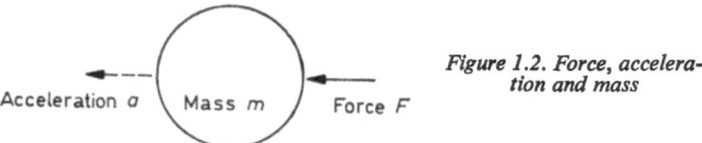

Figure 1.2. Force, acceleration and mass

When *several forces* act on the body simultaneously, the following cases can arise:

2

1. The forces can be formed into a *resultant*, and we may now *replace* the several forces by the one resultant. The *acceleration due to this resultant* will be the same as the combined effect of each of the several forces (*see Figures 1.3(a)* and *1.3(b)*).

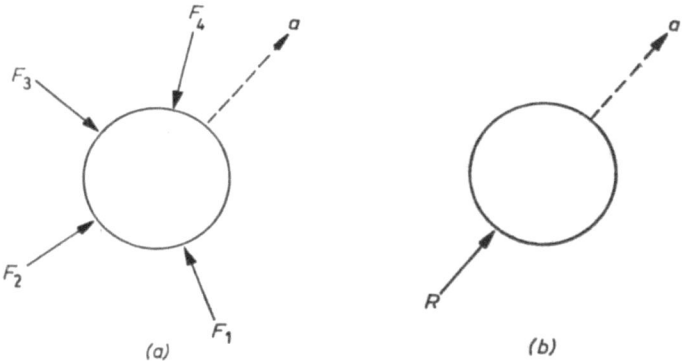

Figure 1.3. (a) Body accelerated under combined effort of F_1, F_2, F_3 *and* F_4. *(b) Body accelerated under effect of single force* R *(resultant) which replaces force system of Figure 1.3(a)*

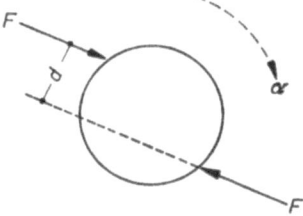

Figure 1.4. Rotational acceleration

2. The forces consist of a *couple*, i.e. *two* forces, which are equal, opposite and at a parallel distance *d*, as shown in *Figure 1.4*.

The couple is termed a *moment* and causes *rotation* of the body. The rotational acceleration a (*angular* acceleration) is proportional to the moment M (where $M = F \times d$) and the constant of proportionality is now the *moment of inertia*, I, of the mass m:

$$M = I.a$$

Note that calculation of the moment of inertia is based on the formula

$$I = \Sigma(k^2.dm) \quad \text{or} \quad I = \int k^2.dm$$

where dm is an element of the mass and, k is its distance from the axis of rotation (usually at the centre of gravity).

3. Generally, *both a resultant force and a moment* can be formed to represent the force system on the body. Then, as shown in *Figure 1.5*, the body will be accelerated both *linearly* and *rotationally*, according to:

Linear acceleration $\qquad a = \dfrac{R}{m}$

Rotational acceleration $\quad a = \dfrac{M}{I}$

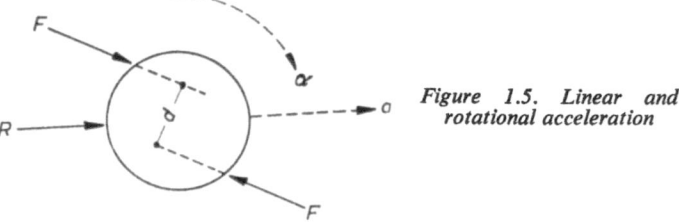

Figure 1.5. Linear and rotational acceleration

4. A special case arises when *both* the resultant force R *and* the moment M are *zero*. Then $a = a = 0$ and the body is either *at rest or moves with uniform velocity*. The body is now considered to be in *equilibrium*. This case is the basis for

4

Newton's First Law which states that a body will be at rest or in motion at uniform velocity *unless a resultant force acts on it*. Thus Newton's First Law is merely a special case of his Second Law: $F = m.a$. In this special case, the resultant force is zero and therefore, the acceleration, proportional to the force, will also be zero.

To show this in figures:

Since in equilibrium $a = a = 0$
$$F = m.a = m.0 = 0 \quad \text{and}$$
$$M = I.a = I.0 = 0$$

That is, the *resultant force and moment are zero when the body is at rest or moving with a uniform velocity*. The concept of equilibrium is illustrated by an example in *Figures 1.6(a)* and (*b*):

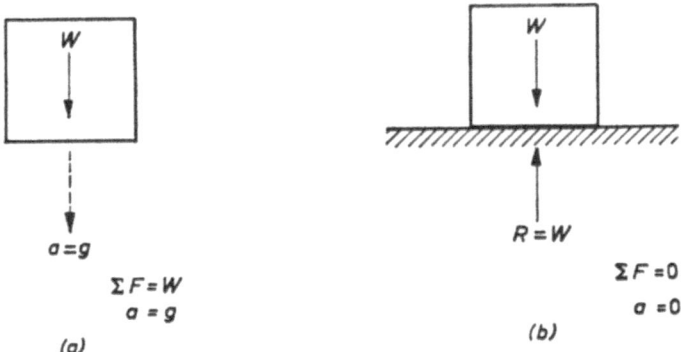

Figure 1.6. (a) Body not in equilibrium. (b) Body in equilibrium

If a body of weight W is released in space, it will fall with the constant acceleration of gravity, g, under the action of the weight force W (as shown in *Figure 1.6(a)*). On the other hand, as shown in *Figure 1.6(b)*, if the body is resting on a supporting surface, it will be acted upon by the reaction force of the surface, equal and opposite to the weight force: thus the resultant force on the body will be zero, the body will be in equilibrium.

Statics deals only with force systems which comply with the requirements of item 4. These are *systems in equilibrium*. From the foregoing discussion it follows that Statics can be regarded as a *limiting* or special case of Dynamics.

1.3. EXTERNAL FORCES

All forces applied *to the body as a whole* are termed external forces. In considering the external forces, we distinguish between active forces or *loads* and passive forces or *reactions*.

Weight is the most common load, it is due to the acceleration caused by the Earth's gravity. The weight force acts vertically, through the centre of gravity of the body.

Tables of specific weight or density list the weight of unit volume of various materials and based on these figures, loads due to weight may be calculated.

In the design of building structures, the weight of the structure itself is termed the *dead load*. To this is added the *applied load* which is an estimate of the weight of people, furniture and stored materials.

An important building load is due to the action of the *wind*. These *wind loads* are assessed in Building Codes; there is a variation in the wind speed according to locality. The empirical formula

$$p = \frac{V^2}{400}$$

allows the calculation of the *wind load p* (lb/ft^2) from the stated or estimated wind speed V (mile/h).

Often weight and wind are the loads on a structure as illustrated by *Figure 1.7* which shows a water tank and its supporting structure.

Wind forces are of dynamic origin since they are due to the

movement of masses of air. Other types of forces are also of dynamic character, such as: *Centrifugal forces* which arise out of rotational motion; *Accelerational and braking forces* which are due to changes in the motion of travelling cranes, trains or machinery and *Impact forces* which are caused by the sudden application of loads. *Earthquake loads* are also dynamic in character.

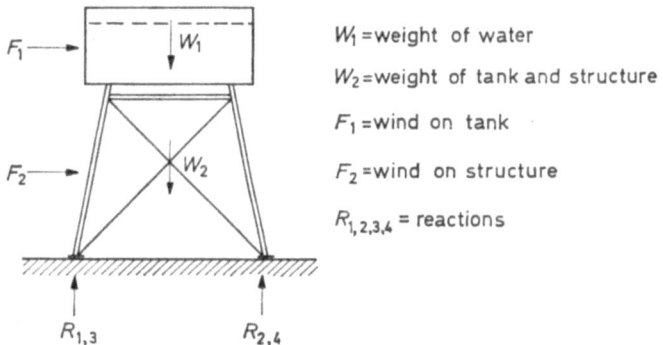

W_1 = weight of water

W_2 = weight of tank and structure

F_1 = wind on tank

F_2 = wind on structure

$R_{1,2,3,4}$ = reactions

Figure 1.7. Elevated water tank

Other forces are due to the inherent material characteristics of the body: *Friction and viscous forces* which are due to molecular properties, and *Expansion and shrinkage forces* which arise out of changes of moisture content, variation in temperature, or chemical action.

In the process of designing a structure or a machine, a list is made of all relevant forces and the location of these loads is indicated on a line diagram.

Almost all structures are firmly supported and therefore, additional support forces or passive forces are acting. These *reactions* must be determined in order to obtain *all* the external forces acting on the structure.

The external forces acting on an aircraft in uniform flight,

7

however, are complete *without fixed support* forces; the aerodynamic lift on the wings replaces the reactions of firm supports (*see Figure 1.8*).

W = total weight of aircraft
L = lift force

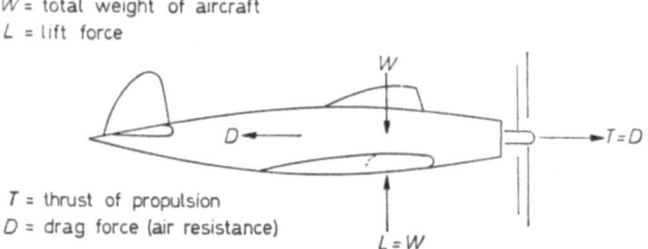

T = thrust of propulsion
D = drag force (air resistance)

Figure 1.8. Aircraft in uniform level flight

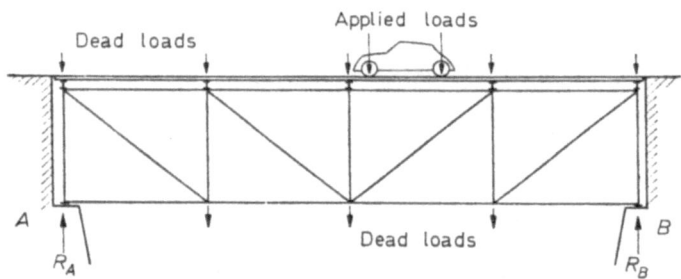

Figure 1.9. Outline of bridge structure

All the external forces, *including loads and reactions*, make up the total force system acting on the body. When the body is at rest or in uniform motion, this total system of forces is in equilibrium.

In cruising flight at uniform velocity, the force system acting on the aircraft of *Figure 1.8* is in equilibrium. As an illustration of this total force system, let us consider the bridge structure shown in *Figure 1.9*.

8

The bridge supports the dead loads due to the weight of the structure and the applied loads imposed by the traffic passing over it. These acting forces are shown as vertical arrows; in addition, there are reactions at *A* and *B*, completing the total system of forces acting on the bridge structure. This system holds the bridge in equilibrium, and it forms the basis for the design of the bridge structure.

On the other hand, the vehicle passing over the bridge is in itself in vertical equilibrium and if we *isolate* it from the structure as shown in *Figure 1.10(a)*, it is seen that its weight is held in equilibrium by the supporting forces provided by the bridge structure, acting on the wheels. The bridge abutment *A* can also be isolated and its force system is similarly found to be in equilibrium: this is composed of the bridge load L_A and the ground reaction R_A, see *Figure 1.10(b)*.

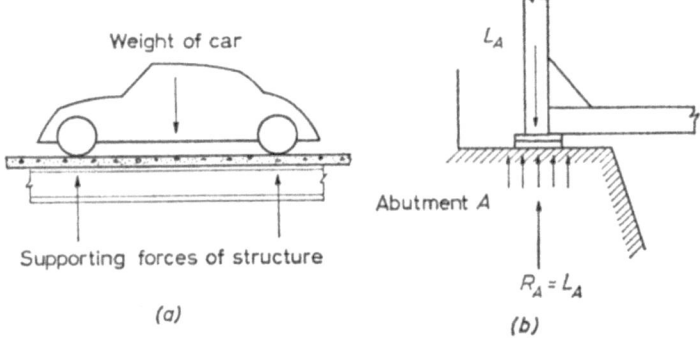

Figure 1.10. (a) Equilibrium of vehicle. (b) Equilibrium of abutment

1.4. INTERNAL FORCES

In resisting deformation by external forces, a body exerts *internal forces*. This resistance to deformation is due to molecular properties and is known as the *strength* of the material.

9

Consider the brick wall built on a concrete base, shown in *Figure 1.11*. The horizontal force F is applied to the top of the wall and the weight W of the wall acts vertically down through its centre.

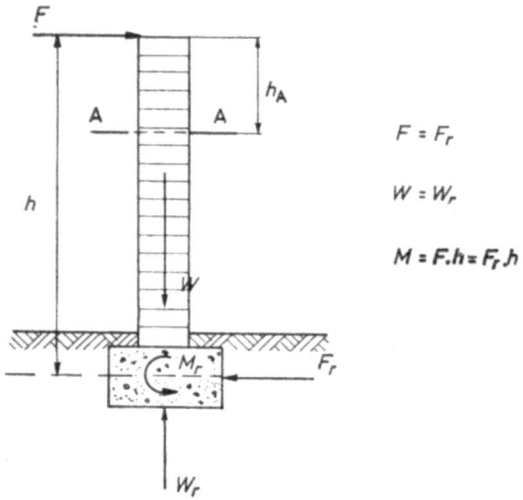

$$F = F_r$$

$$W = W_r$$

$$M = F.h = F_r.h$$

Figure 1.11. Wall on concrete foundation

F and W are applied loads, and equilibrium is completed by F_r, W_r and M_r, the *reactions*, which keep the structure at rest. W_r resists downwards movement due to W; F_r resists sideways movement due to F, and M_r is a moment reaction or couple at the base which resists the overturning couple, M, produced by the combined effect of F and F_r.

Thus the wall is in *equilibrium* under the total force system shown in *Figure 1.11*, as both *resultant* force and moment *are zero*. This presumes that the active load at the *top of the wall*, F, can be resisted by the reaction F_r at the *base*, that is, it has been assumed that the wall is capable of *transferring F* to the

10

base. The transfer of loads through the structure to a safe and resistant foundation is an important aspect of Statics. Note how in *Figure 1.9* the vehicle load is transferred first onto the bridge, then to the abutment structure and finally, to solid ground. In order to understand the mechanism of this transfer, we will use the following reasoning: Let us assume that the wall is cut along line A—A; that part of the wall which is above A—A is reproduced in *Figure 1.12* (the weight of the wall is momentarily neglected).

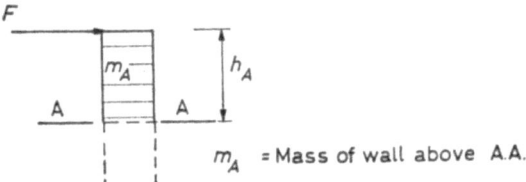

Figure 1.12. Top portion of wall

On this top portion of the wall the force F is the total (resultant) force system and, according to Newton's Second Law:

$$F = m.a$$

From this, $a = \dfrac{F}{m}$. This is not zero, and the top of the wall will slide off the bottom part with the acceleration a.

Rotation will also occur:

$$M = F.h_A = I.a$$
$$a = \frac{F.h_A}{I}$$

which is again not zero.

If we now imagine that the bond of the brickwork across the cut A—A is restored, we must conclude that it is due to its *strength* (or molecular adhesion) across A—A, and indeed, *across every similar section of the wall*, that it can resist the

11

tendency of the upper part to slide or rotate off the bottom part. This resistance is due to the *strength* of the material (brick, mortar), which makes up the wall and is termed the *internal force*. The *transfer* of force *F* to the foundation is only possible if the wall has sufficient strength to resist sliding or rotation in *every* section.

Actually, *F* could be increased to the point where its effect will exceed the molecular bond of the material and then the structure will *fail*. During failure, the structure will behave as described above: it will slide and rotate off its base.

In the following discussion it will be assumed that the material is capable of exerting sufficient resistance to prevent failure. The capacity of the material to resist sliding is called *shearing strength* and the total sliding force exerted along a section is the *shearing* force. Resistance to rotation is termed the *bending strength* of the section and the total couple acting at the section is the *bending moment*.

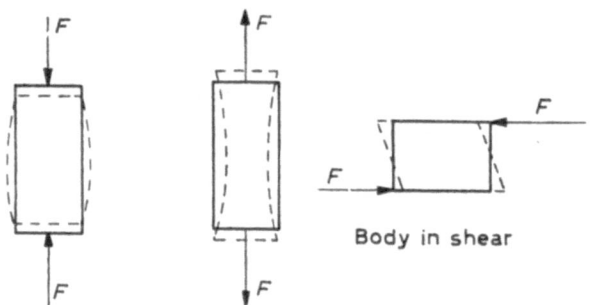

Body in compression   Body in tension

Figure 1.13. Thrust and shear forces

Referring to *Figure 1.12*, the shearing force in Section A—A is *F*, the bending moment is $F.h_A$. If, as in *Figure 1.11*, the weight of the wall is also considered, every section of the wall must resist the crushing action of the wall portions above it. Resistance to crushing is termed *compressive strength* and the crushing force is a *compressive* force, i.e. the material is in *compression*.

Where a force tends to pull the material apart, the force is called a *tensile* force, i.e. the material is in *tension*. Both compressive and tensile forces are termed *thrust* forces. A thrust force is parallel to the axis of the body as distinct from a shear force which is at right angles to the axis. This is shown in *Figure 1.13*.

The shearing force, bending moment and thrust are the *internal forces* in every section of the wall. If the wall is incapable of developing *any one* of these in *any one section*, failure will occur. The designer of a structure must therefore specify such materials and dimensions that under the prescribed external force system, *all internal forces are safely developed*.

Example 1.1

Calculate the shearing force, thrust and bending moment in the

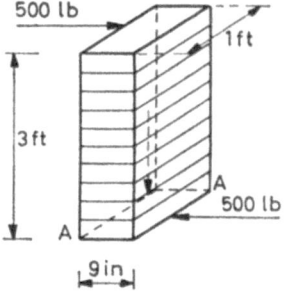

Figure 1.14. Wall portion of Example 1.1

wall shown in *Figure 1.11*, at a section 3 ft from the top. The horizontal force *F* is 500 lb/ft length of wall. *Figure 1.14* reproduces the top portion concerned, showing 1 ft length of wall.

(*a*) Shearing force in the section A—A equals force F:

Shearing force = 500 lb/ft length of wall

(*b*) Thrust in the section is due to the weight of the wall portion above the section and since brickwork weighs 120 lb/ft³:

Thrust = 3 ft $\times \dfrac{9}{12}$ ft $\times$ 1 ft $\times$ 120 lb/ft³ = 270 lb/ft length of wall

(*c*) Bending moment in the section is the rotational effect of force F on the section:

Bending moment = 500 lb $\times$ 3 ft = 1,500 lb-ft/ft length of wall

Note: Generally, internal forces *vary from section to section*. In the case of Example 1.1, the shearing force remains the same along the entire height of the wall, whereas the thrust and bending moment increase with the distance of the section from the top. For loads other than those in Example 1.1, this may no longer hold. In every case the internal forces must be determined for each section separately.

Example 1.2

A tubular aerial mast and the loads acting on it are shown in *Figure 1.15*. Calculate internal forces at (*a*) Section A and (*b*) Section B.

In *Figure 1.16* that portion of the mast above Section A is shown with external forces acting on the portion.

(*a*) Calculation of internal forces at Section A:

Shearing force = 100 lb
Thrust = 200 lb (compression)
Bending moment:
 rotation in plane AOx:

$M_y = 200$ lb $\times$ 2 ft = 400 lb-ft

rotation in plane AOy:

$M_x = 100$ lb $\times$ 4 ft = 400 lb-ft

Twisting moment (rotation of tube around its own axis):

$M_z = 100$ lb $\times$ 1 ft = 100 lb-ft

(b) Calculation of internal forces at Section **B**:

Shearing force $= 300 - 100 = 200$ lb
Thrust $= 200$ lb (compression)

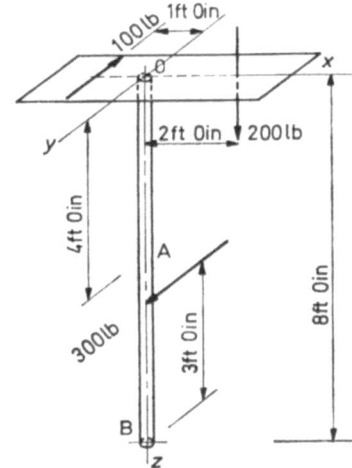

Figure 1.15. Tubular aerial mast of Example 1.2

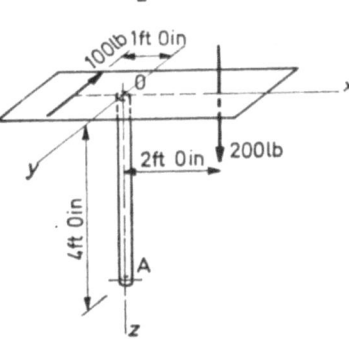

Figure 1.16. Portion of mast above Section A

Bending moment:
rotation in plane A$O x$:

$$M_y = 200 \text{ lb} \times 2 \text{ ft} = 400 \text{ lb-ft}$$

15

rotation in plane AOy:

$$M_x = 300\,\text{lb} \times 3\,\text{ft} - 100\,\text{lb} \times 8\,\text{ft} = 100\,\text{lb-ft}$$

Twisting moment: $M_z = 100\,\text{lb} \times 1\,\text{ft} = 100\,\text{lb-ft}$

Note that the 300 lb force does not enter into the calculations of part (*a*) since the portion above Section A does not carry this load and thus the total force system of this portion is complete without it.

Summing up, the internal forces of *any section* are determined by considering the action of the *total* external force system *at the section*. This will be illustrated in greater detail in Chapter 5.

1.5. CONCENTRATED AND DISTRIBUTED FORCES

Loads may be either concentrated or distributed.

A *concentrated load* is one which acts at a point. In practice, this is rarely the case since a small contact area is required

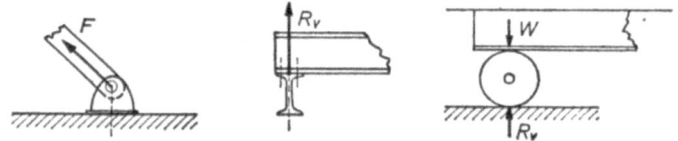

Figure 1.17. Examples of concentrated loads and reactions

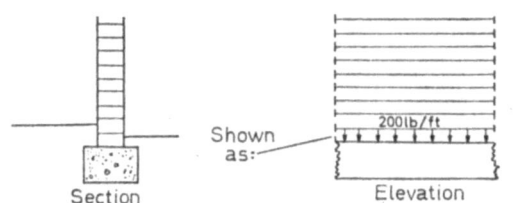

Figure 1.18. Wall load: example of a distributed load

through which the force can act. Forces which are assumed as concentrated are those acting through frame hinges, beam supports or rollers as shown in *Figure 1.17*.

16

Distributed loads are spread over a certain area. Wall loads on footing beams or wind loads on roofs are distributed loads, *see Figure 1.18.*

When equally and closely spaced concentrated loads are acting, they may be replaced by an *equivalent* (but approximated) distributed load system.

Example 1.3

A *rigid frame* structure is shown in *Figure 1.19*. The frame is supporting purlins at 3 ft 6 in centres; each purlin transmits a load of 1,200 lb to the frame.

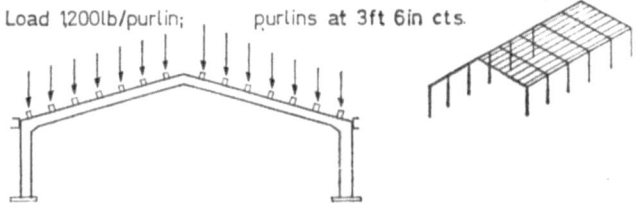

Figure 1.19. The rigid frame of Example 1.3

The equivalent distributed load is

$$w = \frac{1,200 \text{ lb}}{3 \cdot 5 \text{ ft}} = 340 \text{ lb/ft along rafter.}$$

Building Codes specify roof and floor loads *per square foot of area.* These floor and roof loads are converted into distributed loads of the structure.

Example 1.4

The floor structure of *Figure 1.20* consists of beams which span 30 ft across the building and a floor slab which spans 14 ft between beams. The floor slab is supported on the beams and the beams sit on the walls; the floor loads are transferred through slab, beams and walls finally to the wall footings which are supported by the bearing strength of the foundation soil.

17

Load on *concrete slab* (specific weight of concrete is 150 lb/ft³):

$$\text{Weight: } \frac{6}{12} \text{ ft} \times 150 \text{ lb/ft}^3 = 75 \text{ lb/ft}^2$$

Applied load (assumed) $= 40 \text{ lb/ft}^2$

Total design load on slab $= 115 \text{ lb/ft}^2$

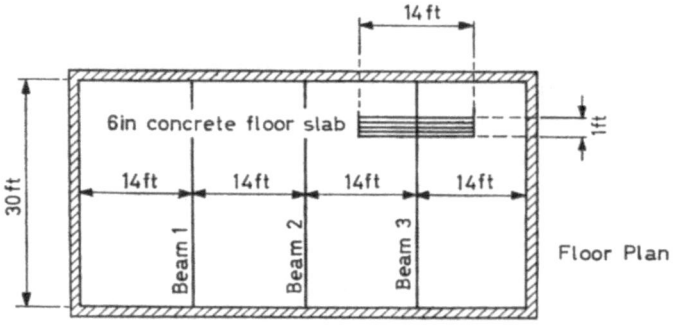

Figure 1.20. Slab and beam floor of Example 1.4

Load on *beams*:
Each foot length of the beams supports 14 ft × 1 ft loaded area (*see* shaded portion in *Figure 1.20*).
Then distributed load on beam:

$$14 \text{ ft} \times 1 \text{ ft} \times 115 \text{ lb/ft}^2 = 1,610 \text{ lb/ft}$$

(Note: this calculation assumes that the slabs are *simply supported* over the beams.)

1.6. PRESSURE AND STRESS

In the mechanics of fluids, *pressure* is defined as the force *exerted on unit area* by a liquid, gas or soil, owing to its weight or its molecular action. A fluid molecule will exert pressure on its neighbouring molecule and on the wall of the vessel.

18

Stress is the *internal force* per square inch of a section of the *solid body*, due to the shear, thrust, bending or twisting moment. Pressure and stress are both in lb/in² units but should not be confused. Both concepts are used when *pressure* of a fluid causes *stress* in the walls of a containing vessel. Pressure is the unit external force in a fluid or soil body, whilst stress is the unit (internal) resisting force in the cross-section of the structure which contains the fluid.

As an illustration, consider the thin-walled, cylindrical pressure vessel shown in *Figure 1.21.*

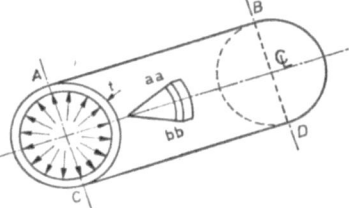

Figure 1.21. Thin walled pressure vessel

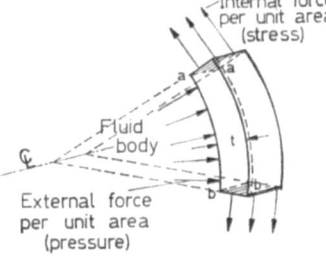

Figure 1.22. Wall element of pressure vessel

A wall element of this cylinder is shown in *Figure 1.22*, and we may consider the equilibrium of this element. The pressure in the cylinder produces the applied or external force on the section; this force is resisted by the internal force or stress in the cylinder wall. To produce equilibrium of the section, the internal force will be in the direction shown, causing tension in the cylinder wall.

19

Internal forces cause stresses according to the way in which the material of the body resists the loading. In a given section, the shearing force induces shearing stresses. Likewise, bending and torsional stresses are induced by the bending moment and twisting moment in the section concerned.

It should be noted that these stresses are normally not uniform throughout a given section of the body. The distribution of stresses is the object of study in Strength of Materials and does not concern us here.

Example 1.5

Find the *bearing pressure* of the brick wall foundation on the soil on which it rests, *see Figure 1.23.*

Specific weight of brickwork: 120 lb/ft³, and of concrete: 150 lb/ft³

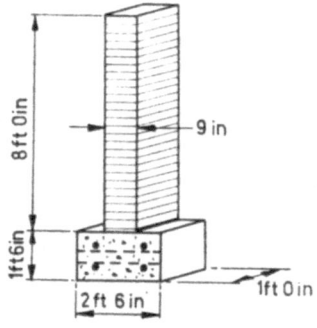

Figure 1.23. Brick wall foundation of Example 1.5

In order to find the total force transmitted from the structure to the soil, both the weight of the wall and of the footing will be calculated.

Considering *1 ft length of wall*:

Volume of brickwork $= 8 \text{ ft} \times \dfrac{9}{12} \text{ ft} \times 1 \text{ ft} = 6{\cdot}0 \text{ ft}^3$

Weight of brickwork $= 6{\cdot}0 \text{ ft}^3 \times 120 \text{ lb/ft}^3 = 720 \text{ lb}$
Volume of concrete $= 2{\cdot}5 \text{ ft} \times 1{\cdot}5 \text{ ft} \times 1 \text{ ft} = 3{\cdot}75 \text{ ft}^3$
Weight of concrete $= 3{\cdot}75 \text{ ft}^3 \times 150 \text{ lb/ft}^3 = 562 \text{ lb}$

20

Force transmitted to soil = 720 lb + 562 lb = 1,282 lb
This force is distributed over an area of 2·5 ft × 1 ft = 2·5 ft²

Thus bearing pressure = $\dfrac{1,282 \text{ lb.}}{2\cdot5 \text{ ft}^2}$ = 512 lb/ft²

Note: It is important to maintain consistent use of units when carrying out calculations. In Example 1.5 it was advantageous to use ft units throughout which allowed direct insertion of the values of density (in lb/ft³ units) and resulted in lb/ft² units for the bearing pressure.

Vessels which contain a liquid or a gas will sustain pressure loads. Gas molecules arrange themselves uniformly around the confined volume of the vessel and it is often assumed that the gas exerts *uniform* pressure on the walls of the container. The molecules of a liquid, on the other hand, being in close contact with each other in the liquid volume, exert their cumulative weight on the confining surfaces. Therefore, the higher the liquid column is, the greater its pressure will be at the base.

It can be shown that the *liquid pressure*, p, at the depth h below the free surface of the liquid is

$$p = w.h$$

where w, the constant of proportionality, is the specific weight of the liquid.

The proof of the formula takes us into the realm of Fluid Statics (Hydrostatics). In *Figure 1.24*, a column of liquid is shown between the free surface and Section A—A at depth h_A.

Figure 1.24. Pressure at the base of a column of liquid

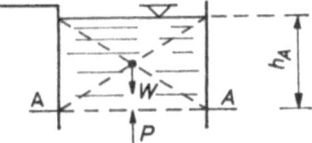

The column of liquid is in equilibrium. In the vertical direction, there are only two forces acting: W, the weight of the liquid column, and P, the total *pressure force* at the level A—A.

21

$$P = W$$

Also,
$$W = w.h_A.A$$

where w is the specific weight of the liquid and A is the cross-sectional area at A—A.

But
$$P = p.A, \text{ which means that}$$
$$p.A = w.h_A.A$$

That is,
$$p = w.h_A$$

Since the pressure is the same everywhere on a horizontal plane, the value of pressure p is the same on the wall of the vessel at this depth.

Example 1.6

Figure 1.25 shows the wall of a reservoir which is holding back a volume of water. What is the pressure:

(a) at 10 ft depth;
(b) at the bottom of the reservoir?

Specific weight of water is 62·5 lb/ft³

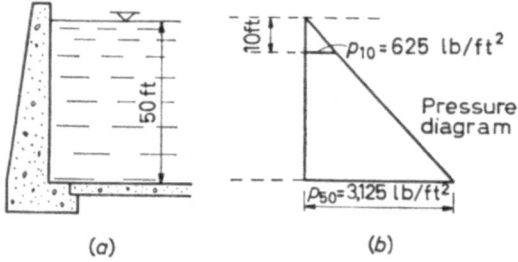

(a) (b)

Figure 1.25. Reservoir of Example 1.6

(a) At 10 ft depth the pressure

$$p_{10} = w.10 = 625 \text{ lb/ft}^2$$

(b) At 50 ft depth:

$$p_{50} = w.50 = 3{,}125 \text{ lb/ft}^2$$

In *Figure 1.25(b)* a *pressure diagram* shows the linear variation of

22

pressure with depth. Often we are interested in the *total force* due to pressure: this is the *resultant pressure force* and it is obtained by forming a summation of all the elementary unit pressures on a section of the wall.

If in Example 1.6 we were to calculate the resultant pressure force acting on 1 ft width of wall (the 1 ft taken at right angles to the paper), this would be the way of doing it:

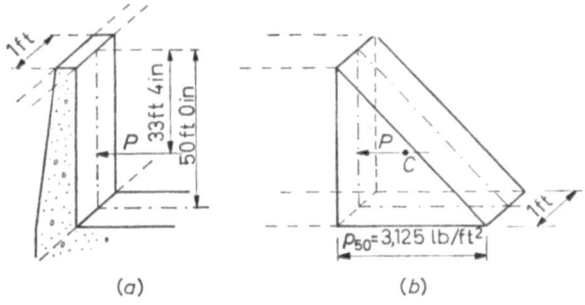

Figure 1.26. Calculation of the resultant pressure force

Since the pressure varies over the surface 50 ft deep × 1 ft wide, a summation could be made by taking each individual pressure, multiplying it with the area over which it acts (in our case, 1 ft²) and then adding up these incremental forces. But we can do the same by taking the *average* pressure and then multiplying it with the *total* area of the wall.

Average pressure: $\dfrac{0 + 3,125}{2} = 1,562$ lb/ft²

Wall area: $50 \times 1 = 50$ ft²

Resultant pressure force: $1,562 \times 50 = 78,100$ lb

Or, generally, on b ft width of wall:

Average pressure: $\dfrac{0 + w.h}{2} = \dfrac{w.h}{2}$

Wall area: $b.h$

Resultant pressure force: $\dfrac{w.b.h^2}{2}$

23

The line of the resultant will be through the centroid of the pressure diagram, that is, one-third the depth up from the base.

This completes Example 1.6.

1.7. SUMMARY: BASIC CONCEPTS

Forces and moments are external influences on rigid bodies. The acting loads and support reactions together may form a *system of forces which is in equilibrium.* We regard this as an extreme or special case of Dynamics when the acceleration is zero, and deduce that this can only occur when the resultant of acting forces is zero.

External forces may be concentrated or distributed. Liquid pressures cause a loading which varies in a linear manner with depth. Wind pressures depend on the square of the wind speed.

Internal forces are induced by the action of external forces; they represent the resistance of the body to deformation or failure. There are shearing and thrust forces and bending and twisting moments.

Stresses are internal forces over unit sectional area of the body. There are shearing, thrust, bending or torsional stresses and they will generally vary throughout the section.

1.8. PROBLEMS

Problem 1.1

Calculate the design load on a floor which consists of a 6 in thick reinforced concrete slab, lined on its underside with 1 in thick insulating material (specific weight 60 lb/ft³). Applied load is 50 lb/ft²

ANS. 130 lb/ft²

Problem 1.2

What force is produced at the wall in holding the bracket of *Figure 1.27* in equilibrium?

ANS. 500 lb

Problem 1.3

A street lamp is rigidly encased into the pavement at *A*, as shown in *Figure 1.27a*. Find the shearing force, thrust and bending moment at *A* for the following system of loading:

Weight of lamp W_1 = 110 lb
Weight of arm W_2 = 140 lb
Weight of post W_3 = 300 lb
Wind load on structure W_4 = 160 lb

ANS. Shearing force 160 lb
Thrust 550 lb, compression
Bending moment 2,680 lb-ft

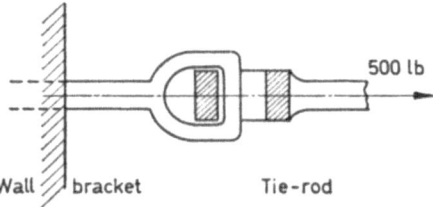

500 lb

Wall bracket   Tie-rod

Figure 1.27. Bracket and tie-rod of Problem 1.2

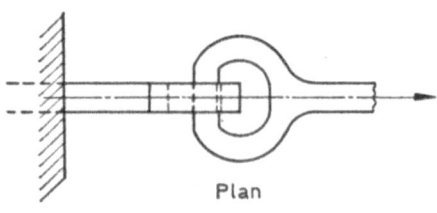

Plan

Figure 1.27a. The street lamp of Problem 1.3

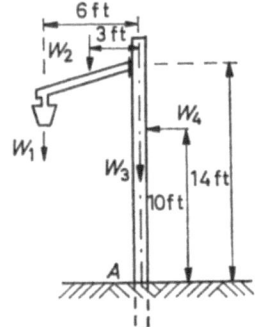

Problem 1.4

The exhaust valve of an internal combustion engine is shown in

Figure 1.28. The spring force in the closed position as shown is 20 lb and the pressure in the cylinder is 30 lb/in² above that in the exhaust port.

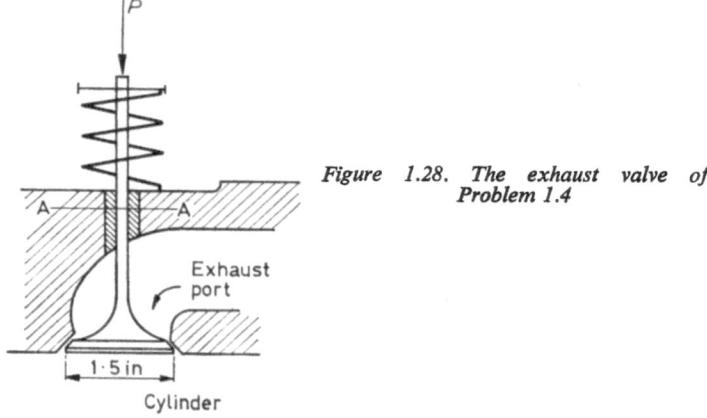

Figure 1.28. The exhaust valve of Problem 1.4

Find the force P to be applied to the valve stem to cause the valve to open, also the internal force in the valve stem at Section A—A (Neglect weight of valve).

ANS. $P = 73.0$ lb
$F_{A—A} = 53.0$ lb comp.

Problem 1.5

The piston of the air compressor shown in *Figure 1.29* has a number of external forces acting on it due to its weight, the air

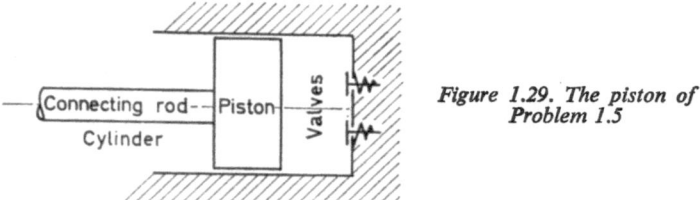

Figure 1.29. The piston of Problem 1.5

pressure in the cylinder, friction between piston and walls and the force transmitted through the connecting rod. Indicate these forces on a diagram.

26

Problem 1.6

The structure for a road overpass is shown in *Figure 1.30*; the *applied* load on its *deck* is 200 lb/ft².

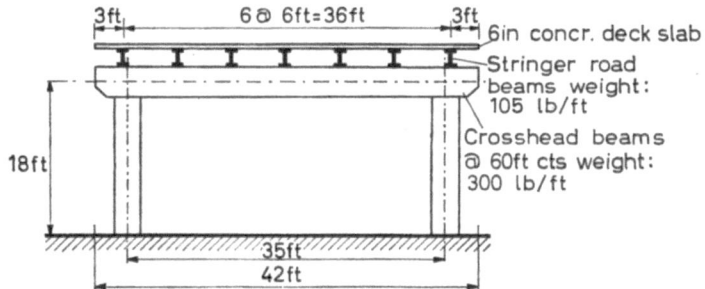

Figure 1.30. The road overpass of Problem 1.6

Find the distributed loading on the stringer road beams and the *equivalent* distributed loading on the crosshead support beams which are 60 ft apart.

ANS. 1,755 lb/ft,   17,850 lb/ft

Problem 1.7

A cable which is made up of seven equal and parallel strands, supports a load of 5 tons. Find the magnitude and nature of the internal force in each strand.

ANS. 1,600 lb tension

Problem 1.8

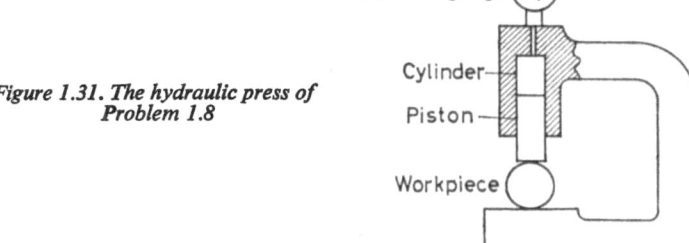

Figure 1.31. The hydraulic press of Problem 1.8

The hydraulic press shown in *Figure 1.31* applies a force to the workpiece through a 2 in diameter piston. If the pressure gauge reads 1,000 lb/in², calculate the force applied to the workpiece.

ANS. 3,140 lb

Problem 1.9

The three weights in *Figure 1.32* are supported on three connecting cables *A*, *B* and *C*. What are the internal forces in the cables?

ANS. *A* = 6 tons, *B* = 3 tons, *C* = 1 ton (all in tension)

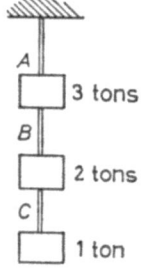

Figure 1.32. Suspended weights of Problem 1.9

Problem 1.10

A winch lifts a 100 lb weight at constant speed by means of the

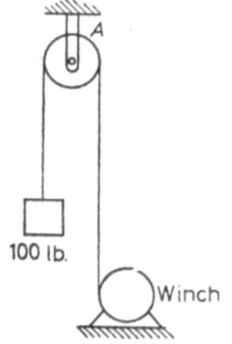

Figure 1.33. Winch and pulley of Problem 1.10

pulley in *Figure 1.33*. What force is applied to the pulley support bracket at *A*?

ANS. 200 lb tension

Problem 1.11

Determine the *total* force (resultant pressure force) acting on the wall of the dam shown in *Figure 1.34*. The dam wall is 100 ft long. Specific weight of water = 62·5 lb/ft³.

ANS. 3,480 tons

Figure 1.34. Dam of Problem 1.11

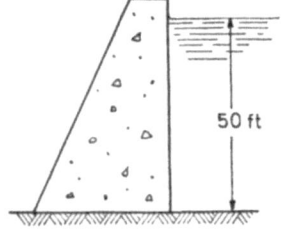

Problem 1.12

Figure 1.35 shows the cross-section of a *composite* beam consisting of a steel beam encased in concrete. Find the weight per foot length of the beam if:

Cross-sectional area of steel beam = 15 in²
Specific weight of steel = 0·28 lb/in³
Specific weight of concrete = 140 lb/ft³

ANS. 230 lb/ft

Figure 1.35. Composite beam of Problem 1.12

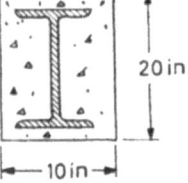

29

CHAPTER 2

MANIPULATION OF FORCES

2.1. Forces are vectors

Scalars are physical quantities which have *magnitude* only. Such quantities are length, time and mass.

Vectors are physical quantities which have *direction* as well as magnitude: velocity, acceleration, momentum and force are vector quantities.

In Chapter 1 we have seen that acceleration and force are closely related; *force* causes an *acceleration* to which it is proportional:

$$F = m \times a$$

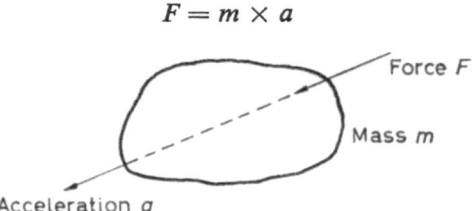

Figure 2.1. Force and acceleration

The mass *m* is a scalar, a simple multiplying factor. *F* and *a* are vectors and since the above equation is correct both in *magnitude and direction*, the directions of *F* and *a* must be identical. It could also be said that it is a *vector equation* and

vector $F = m \times$ vector a

Force vectors can be *free* or *fixed*. A free vector is one which can be considered as acting at any point along its line of action. On the other hand, a fixed vector acts at a specific point.

Figure 2.2 shows a beam supported at *A* and *B* and loaded with forces F_C and F_D. It is convenient to assume that support forces F_A and F_B and support moment M_B are acting *at* the support points *A* and *B*. Similarly, loads F_C and F_D are acting *on* the beam at points *C* and *D*.

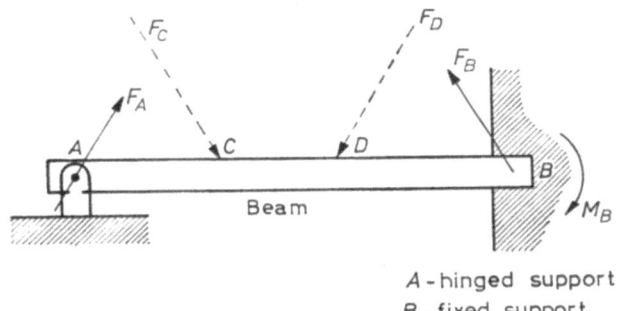

A - hinged support
B - fixed support

Figure 2.2. Free and fixed forces

For most purposes, these forces may be *freed* and moved along their line of action to whichever point is most convenient for manipulation in Statics.

Force systems tend to be complex: even a few forces scattered in space are likely to make it difficult to determine whether they are in equilibrium. Often it is essential to manipulate external forces and to form a *resultant*. Then it is the resultant (which has, in the general case, six components) that is acting on the body and if we wish to establish equilibrium, we must ensure that a *balancing force* opposes the resultant.

Chapter 2 leads to systematic methods used to form a resultant and to the concept of the balancing force.

Manipulation of vectors can be carried out in two ways: by *graphical* means or by *calculation*. If we wish to apply the principles of graphics, let us use a well-sharpened pencil, an inch rule and two set-squares (an india-rubber will surely be

31

handy too). When tackling a problem by calculation, drawing instruments would still be required in order to prepare an outline drawing of the structure and forces; in addition, a slide rule will be essential. Once the use of a slide rule has been mastered, it will greatly speed calculations (there are a number of excellent booklets on slide rules).

Often graphics and calculations can be used in a complementary fashion; since basic principles are identical in the two methods, part of a problem can be solved by graphical means, another part by calculation. If both techniques are familiar, our judgement can be used in selecting the best (fastest and safest) method for a given problem.

2.2. TWO FORCE VECTORS IN A PLANE

If two vectors are in the one plane, the two forces are *co-planar* and the problem is *two-dimensional*.

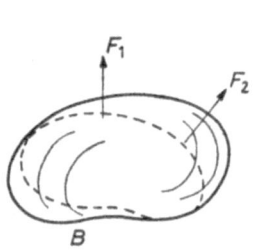

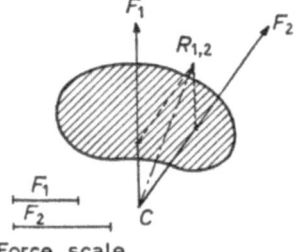

Force scale

Figure 2.3. Two co-planar forces

Figure 2.4. Resultant of two co-planar forces

In *Figure 2.3*, F_1 and F_2 are in the plane of the drawing. This plane of F_1 and F_2 cuts the three-dimensional body B along the dotted line (trace). In *Figure 2.4* the outline of the body B has been omitted and only this trace-line is shown. The free vectors F_1 and F_2 have been moved along to their point of intersection C.

32

The operation of adding F_1 to F_2 has been performed graphically in *Figure 2.4*. A *force scale* was chosen so that certain lengths represent the magnitude of F_1 and of F_2. These lengths were then plotted along F_1 and F_2, respectively, commencing from C.

The diagonal $R_{1,2}$ then, is the resultant of F_1 and F_2:

$$R_{1,2} = F_1 + F_2$$

This is a vector addition equation.

It is usual to perform the graphical addition on a separate diagram as shown in *Figure 2.5*:

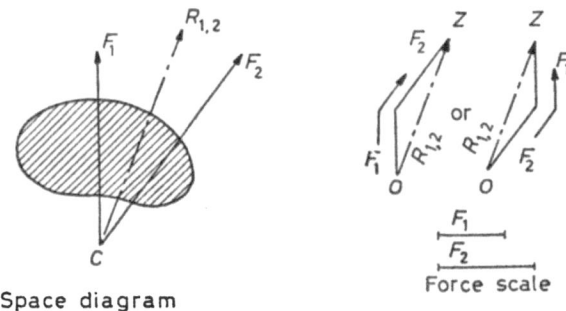

(a) Space diagram

(b) Force diagram

Figure 2.5. Addition of two co-planar forces

On *Figure 2.5(a)* Space diagram, only *positions* of the force vectors are shown and the manipulation of vectors is carried out in *Figure 2.5(b)*, Force diagram. Here, the two forces are plotted in succession (head to tail), parallel to their respective directions in the Space diagram. The resultant, $R_{1,2}$, is obtained by connecting the starting point O to the finishing point Z. The force diagram gives the *magnitude* (read from the force scale) and *direction* of $R_{1,2}$ but its position must be found in the Space diagram. If a parallel to $R_{1,2}$ is drawn through point C in the Space diagram, this gives its position on the body.

33

Naturally, $R_{1,2}$ is not an actual acting force on body B but it has the combined effect of F_1 and F_2.

The equation

$$D_{1,2} = F_1 - F_2$$

is a vector *subtraction* equation and $D_{1,2}$ is the *difference vector* of the system. Vector subtraction can be carried out as an addition if we interpret the $-$ sign of F_2 as a reversal of direction:

$$D_{1,2} = F_1 + (-F_2)$$

That is, F_1 is added to the *reverse* of F_2, and $D_{1,2}$ is *the resultant* of F_1 and $-F_2$. This vector *addition* is shown in *Figure 2.6*:

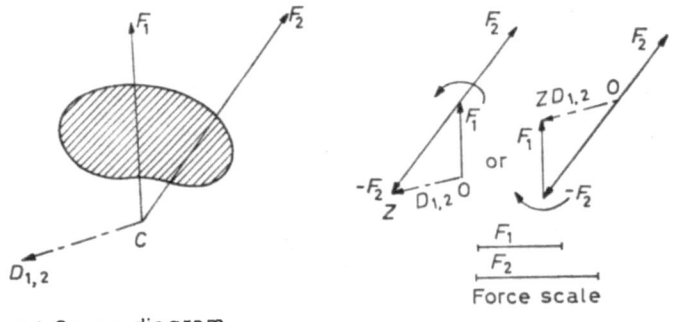

(a) Space diagram

(b) Force diagram

Figure 2.6. Vector subtraction carried out as addition

Note that the graphical construction of *Figure 2.6* is similar to *Figure 2.5* except that the sign (direction) of F_2 has been reversed to $-F_2$. Otherwise, vectors F_1 and $-F_2$ were plotted head to tail and $D_{1,2}$ was obtained by connecting the starting point O to the finishing point Z.

2.3. Two vectors in space

When the two vectors are not in the one plane, this simple method of vector combination is unsuitable since the plane of the drawing cannot contain both F_1 and F_2. *Figure 2.7* shows a space force system of F_1 and F_2; points of intersection of these forces with the co-ordinate planes are D, E and G, H respectively.

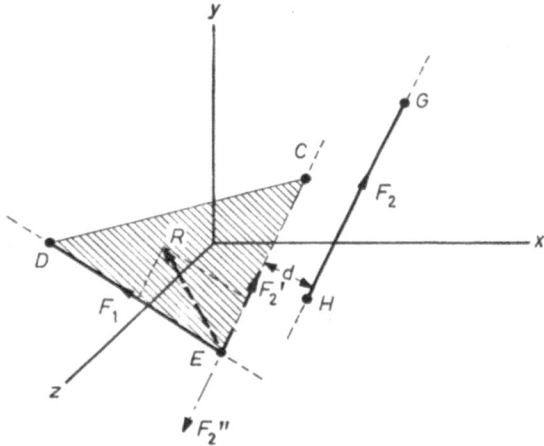

Figure 2.7. Two forces in space

To combine forces F_1 and F_2, the *principle of transposition* will be used. By moving parallel to itself (transposing) the force F_2 *to a point on force* F_1, say to point E, a combination of the two forces will be made possible.

Transposition makes use of an *auxiliary force system* which *in itself* is in equilibrium. Adding this auxiliary system to the system of forces F_1 and F_2, the total value of the latter remains unchanged.

In *Figure 2.7* we added, at point E, the auxiliary system of F_2' and F_2'', both parallel and equal in magnitude to F_2. Since F_2' and F_2'' are equal and opposite, their addition has not changed our original system of F_1 and F_2.

35

F_1 and F_2' are in the same plane (shown shaded in *Figure 2.7*) and can be combined, by the method shown in *Figure 2.5*, into the resultant R which lies in plane $F_1 F_2'$.

Adding the two auxiliary forces F_2' and F_2'', the force system has grown from the original two forces to *four* forces. Two of these have just been combined into the resultant R; the remaining two, namely F_2 and F_2'', form a *couple*. If the perpendicular distance between these two forces is d (the arm of the couple), then the

$$\text{couple or moment} = F_2 . d = F_2'' . d$$

Thus the combination of the two forces F_1 and F_2 which are not in the one plane results in

1. *a resultant force R* (formed from the combination of F_1 and F_2') and

2. *a resultant couple* $F_2 . d = F_2'' . d$ (formed from F_2 and F_2''). The auxiliary system of F_2' and F_2'' is not really needed except to explain the operation.

It should be noted that the value of the resultant force and couple (as shown in *Figure 2.7*) depends on the choice of the point for transposition (in our case, point E). Choosing another point for transposition on force F_1, another, equivalent set of values would be obtained for resultant force and couple. We could have also chosen another auxiliary force system, e.g. F_1' and F_1'' through point H; in this case, F_1' and F_2 would have formed the resultant force and F_1 and F_1'' the resultant couple.

In all these solutions, the arm d is the perpendicular distance from the original position to the transposed position of the force.

It is common practice to represent a couple by a vector whose *magnitude* is $F.d$ (force.arm) and whose *direction* is *at right angles to the plane of the couple*. The force system F_1, F_2 of *Figure 2.7* is then equivalent to two vectors R and M where $M = F_2 . d$ and is at right angles to plane $ECHG$. The body of mass m and moment of inertia I, which has its centre of mass at

E and on which this system acts, is moving forward, in the direction of R, with an acceleration

$a = \dfrac{R}{m}$; simultaneously, it is rotating, parallel to plane

$ECHG$, with an angular acceleration

$a = \dfrac{M}{I}$, so that a spiral motion results.

When two forces are equal and opposite and act along the same line, the resultant is zero and the system is in equilibrium. Obviously, *two forces* which are *not in the same plane*, cannot be in equilibrium.

2.4. THREE FORCE VECTORS IN ONE PLANE

Figure 2.8 shows *three* co-planar forces acting in the plane of the drawing.

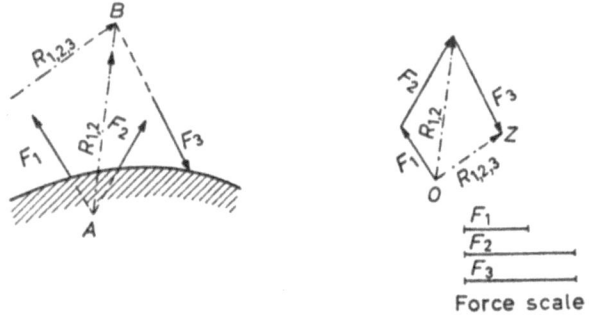

(a) Space diagram   (b) Force diagram

Figure 2.8. Combination of three co-planar forces

In *Figure 2.8(a)* first F_1 and F_2 are combined into a partial resultant $R_{1,2}$ which will pass through point A. The next step is to add F_3 to $R_{1,2}$ and this gives the final resultant $R_{1,2,3}$. It is convenient to plot the three forces to a force scale as shown in

37

Figure 2.8(b), head to tail. First, direction and magnitude of $R_{1,2}$ is obtained and we draw a parallel line to it in *Figure 2.8(a)*; next, this line is brought into intersection with F_3 and through this point B a parallel is drawn to the *direction* of $R_{1,2,3}$ obtained from *Figure 2.8(b)*.

Note that the sequence of plotting the forces in *Figure 2.8(b)* is arbitrary: we may start with any of the three forces and plot them in any sequence. The plotting, however, must be continuous (head to tail) and the forces must be plotted in their proper direction and sense. The resultant will always be given in magnitude, direction and sense as the distance between the starting point O and the end-point Z, measured on the force scale. Whilst the Force diagram supplies $R_{1,2,3}$ in this manner, its actual *position* in the plane of the forces cannot be found directly unless we proceed by the stepwise method outlined above: adding the forces by pairs, forming a partial resultant, adding another force and so on.

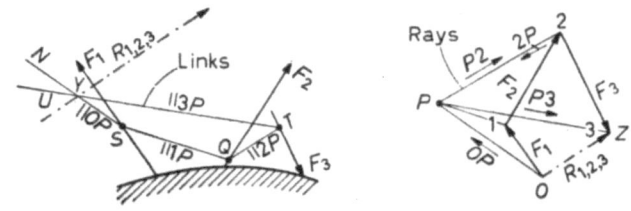

Force scale
as in *Figure 2.8*

(*a*) Space diagram   (*b*) Force diagram

Figure 2.9. Combination of forces by link polygon

The *link polygon* method overcomes this difficulty. In order to explain this method, the three forces were reproduced in *Figure 2.9(a)* and plotted again as before, in *Figure 2.9(b)*, head to tail.

In *Figure 2.9(b)* an arbitrary point, *P*, the pole was chosen. This pole was then connected, in order, to all intersections of the forces and these connecting lines are called *rays*. The first ray is OP; note that O is the starting point of force F_1 on the Force diagram. This ray is next transferred to the Space diagram of *Figure 2.9(a)* by drawing a parallel to it through an *arbitrary point* (say, point S) which is on the line of action of F_1: the resulting line is *link NS*. The graphical construction continues by transferring consecutive rays from the Force diagram to the Space diagram: from point S a parallel is drawn to ray $1P$ and this intersects force F_2 at point Q. From point Q we draw a parallel to ray $2P$ which intersects F_3 at point T and through this the final parallel $3P$ is then drawn, this is link TU. The *intersection of the first link NS and the last link TU* is a point through which $R_{1,2,3}$ passes, this point of intersection is Y. If $R_{1,2,3}$ is found from the force diagram by connecting O to Z, a parallel to it through point Y will give its position and direction in space.

The system of links $NSQTU$ is a link polygon. The starting point for its construction, S, is arbitrary, but once chosen, the system of links is drawn by following a strict sequence. The *link* connecting F_1 and F_2 in the Space diagram, link SQ, must be parallel to the ray which connects P to 1 which is the point *between* vectors F_1 and F_2 in the Force diagram.

To form the resultant, the link polygon uses a system of auxiliary forces: these forces are the rays emanating from the pole. It will be assumed that each of the forces F_1, F_2 and F_3 is resolved into components *along the rays*. Thus, F_1 is the resultant of OP and $P1$, F_2 is the resultant of $1P$ and $P2$ and F_3 is the resultant of $2P$ and $P3$. Then

$$R_{1,2,3} = F_1 + F_2 + F_3 = OP + P1 + 1P + P2 + 2P + P3$$

In this *vector equation*, OP is the vector directed from O to P, and similarly to the other vectors. It is obvious that $P1$ and $1P$ mutually cancel out, also $P2$ and $2P$.

Finally,

$$R_{1,2,3} = OP + P3.$$

This vector equation is represented by the triangle $OP3$ in the Force diagram, *Figure 2.9(b)*. The corresponding forces in the Space diagram are NS (for OP) and TU (for $P3$) and their resultant, $R_{1,2,3}$ would obviously pass through their point of intersection, which is marked as Y.

Choice of the position of the pole P depends only on the convenience of obtaining a link polygon with clear intersections on the force lines, and on our ability to keep the links within the area of the drawing sheet. It occasionally occurs that links become lost towards outer space and then it is time to re-locate P and repeat the construction of the link polygon. The pole can be *anywhere* in the force diagram as long as it yields a convenient and reasonably accurate link polygon.

Example 2.1

Figure 2.10 shows a roof truss with loads acting at panel points.

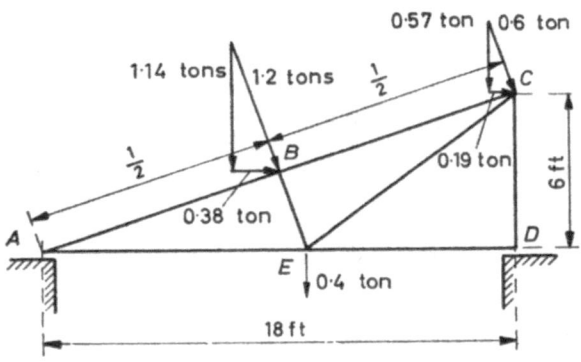

Figure 2.10. Outline of roof truss

(*a*) Calculate the resultant of the loads, including its magnitude, the angle it makes with the horizontal and the distance of its point of intersection with the bottom member of the truss from point A.

(*b*) Find the resultant by means of *link polygon* construction and check on the quantities required in (*a*).

40

(a) For the purpose of *calculating* the resultant, we will use the fact that the *moment of the resultant* about any point is equal to the *sum of the moment* of all the acting forces.

It is convenient to resolve loads at B and C into their horizontal and vertical components.

$$\text{Length } AC = \sqrt{18^2 + 6^2} = 18 \cdot 96 \text{ ft}$$

From similar triangles, horizontal component of load at

$$B = \frac{6}{18 \cdot 96} \times 1 \cdot 2 = 0 \cdot 38 \text{ ton}$$

Similarly, the other load components:

	Load	Vertical component	Horizonta component
B	1·2	1·14	0·38
C	0·6	0·57	0·19
E	0·4	0·40	0
Totals:		$R_v = 2 \cdot 11$	$R_h = 0 \cdot 57$ ton

The position of each of R_v and R_h will now be found separately by taking moments about point A; first, of all the vertical components, and then, of the horizontal components:

$1 \cdot 14 \times 9 \text{ ft} + 0 \cdot 57 \times 18 \text{ ft} + 0 \cdot 4 \times 9 \cdot 9 \text{ ft} = R_v . x \text{ ft} = 2 \cdot 11 . x \text{ ft}$
where x is the distance of R_v from A.
Then $x = 11 \cdot 6$ ft.

Moments of horizontal components about A

$0 \cdot 38 \times 3 \text{ ft} + 0 \cdot 19 \times 6 \text{ ft} = R_h \times y \text{ ft} = 0 \cdot 57 \times y \text{ ft}$
where y is the vertical distance of R_h from A, and $y = 4$ ft.

The resultant R will go through a point 11·6 ft to the right and 4 ft above A and its magnitude is

$$R = \sqrt{R_v^2 + R_h^2} = \sqrt{2 \cdot 11^2 + 0 \cdot 57^2} = 2 \cdot 18 \text{ tons}$$

at an angle

$$a = \tan^{-1} \frac{2 \cdot 11}{0 \cdot 57} = 75 \text{ degrees to the horizontal.}$$

Figure 2.11 shows the position and direction of R as plotted from the *calculations*.

R intersects line AD at a point which is at a distance of $11 \cdot 6$ ft $+ z$ from A, where

$$z = \frac{0 \cdot 57}{2 \cdot 11} \times 4 \text{ ft} = 1 \cdot 1 \text{ ft}$$

The distance from A then is $11 \cdot 6 + 1 \cdot 1 = 12 \cdot 7$ ft.

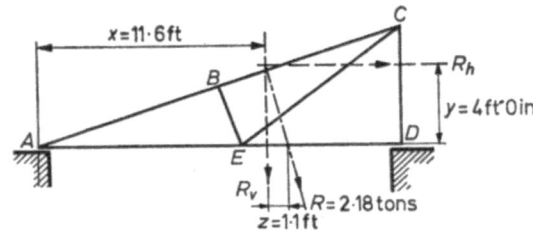

Figure 2.11. Calculated results of example

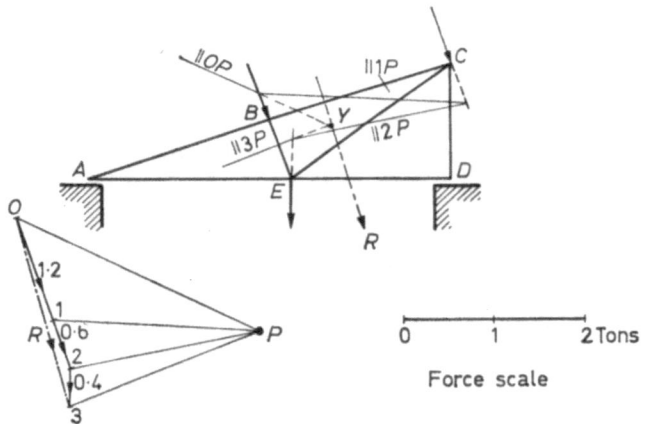

Figure 2.12. Graphical construction of resultant

(*b*) *Graphical solution* of the same example is shown in *Figure 2.12*. The student should verify that these results are the same as those obtained in (*a*).

The graphical method has much to recommend it: it is clear,

42

simple and fast. Although accuracy of graphics is limited, statical calculations rarely require great accuracy.

This completes Example 2.1.

A special case occurs when the three co-planar forces are in *equilibrium*. This can only happen when

(*a*) the three forces, plotted head to tail, form a *closed* polygon (i.e. the head of the last plotted force coincides with the tail of the first) and

(*b*) all three forces intersect at the one point, i.e. they are *concurrent*.

Condition (*a*) simply means that $R_{1,2,3}$ is zero. However, the body on which the three forces act (whilst not accelerating linearly) may still be rotating. *Figure 2.13* shows three forces F_1, F_2 and F_3 which do not intersect at the one point.

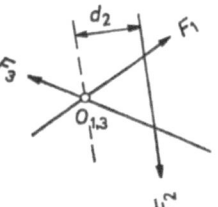

Figure 2.13. Three forces which cannot be in equilibrium

The total rotational moment of the force system can be found by taking moments about, say, $O_{1,3}$. Then $M_{1,3} = F_2 . d_2$. Note that F_1 and F_3 pass through $O_{1,3}$ and their moments about the point are zero. If F_2 were also acting through $O_{1,3}$, the total moment would be reduced to nought, resulting in rotational equilibrium.

Conditions of equilibrium will be discussed in greater detail in Chapter 3.

2.5. THREE VECTORS IN SPACE

When the three forces to be combined are *not in the one plane*, we can extend the method used for two forces and previously shown in *Figure 2.7*. Let us take F_1, F_2 and F_3 as shown in *Figure 2.14*; this force system contains F_1 and F_2 as in *Figure 2.7* and, in addition, a force F_3.

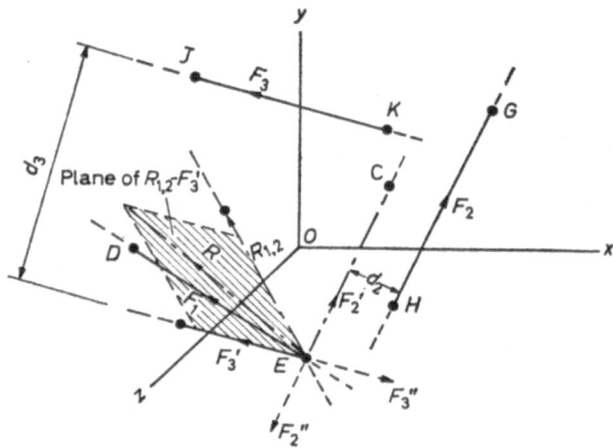

Figure 2.14. Combination of three forces in space

By transposition of F_2 to E, F_1 will be first combined with F_2, resulting as previously in $R_{1,2}$ and $M_2 = F_2 . d_2$. Next F_3 will be transposed to E and from the combination of $R_{1,2}$ and $F_3{}'$ we obtain the total resultant R, and the couple F_3, $F_3{}''$ gives the moment $M_3 = F_3 . d_3$.
Summarizing:

R—is the overall resultant *force*, lying in the $R_{1,2} - F_3{}'$ plane

M_2—a *moment*, a vector at E at right angles to plane $F_2 - F_2{}''$ and

M_3—a *moment*, a vector at E at right angles to plane $F_3 - F_3{}''$.

A further step could be the combination of vectors M_2 and M_3 into a resultant vector M. The resultant rotation at E will then be in a plane at right angles to M.

44

Example 2.2

Three forces are acting on a cubical body of 4 in long sides. Combine these forces to obtain a resultant moment and force at A, *see Figure 2.15.*

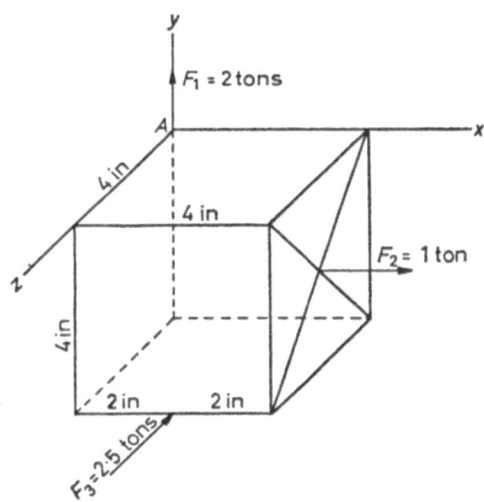

Figure 2.15. Example of force system on cube

(1) Transpose the $F_2 = 1$ ton load to point A, as shown in *Figure 2.16*

$R_{1,2} = \sqrt{4 + 1} = 2.24$ tons, in plane $ABCD$
$d_2 \quad = 2$ in $\times \sqrt{2} = 2.82$ in
$M_2 \quad = 1$ ton $\times 2.82$ in $= 2.82$ tons-in, right angles to AG, in plane $AEFD$.

(2) Transpose the $F_3 = 2.5$ tons load to point A:

$R \quad = \sqrt{4 + 1 + 6.25} = 3.35$ tons in plane $R_{1,2}$—E
$d_3 \quad = \sqrt{4 + 16} = 4.5$ in
$M_3 = 2.5$ tons $\times 4.5$ in $= 11.2$ tons-in, right angles to AH, in plane $ABCD$.

45

Figure 2.17 shows the combination of M_2 and M_3 into a resultant moment at point A, in a diagrammatic manner. In order to obtain the resultant numerically, the vector addition should be carried out in the plane M_2—M_3.

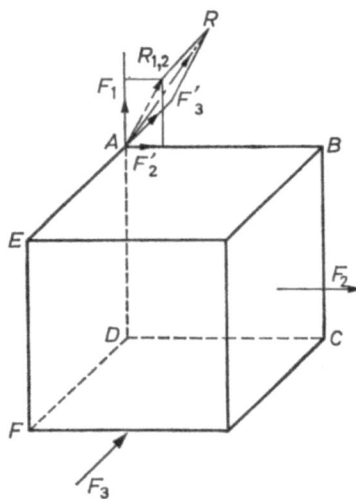

Figure 2.16. Combination of forces, Example 2.2

For the sign of the moment vectors, the following sign convention was adopted: looking down along the vector towards its origin (in our case, A) the vector is directed *towards us* if the moment *is clockwise*. Otherwise, if we observe the moment anti-clockwise, the arrow of the moment-vector will point towards the origin, that is, A.

Summing up, the three forces acting on the cube were combined into a single resultant force R and a single resultant moment M, both acting at A. If the cube were free to move, point A would translate along the line of action of R and rotate in a plane at right angles to M, with appropriate accelerations.

This completes Example 2.2.

This method of step-by-step combination can become tedious when angles are not as regular as above. A methodical way to combine forces will now be explained with reference to a greater number of forces and this method will be applicable to *any number of forces* acting: two, three or more.

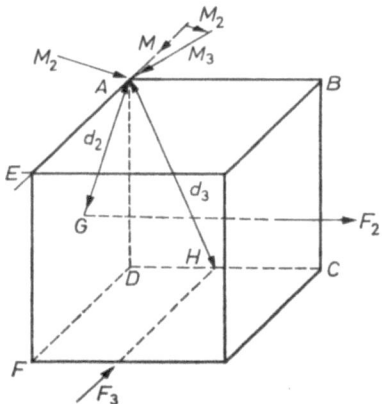

Figure 2.17. Combination of moments, Example 2.2

2.6. FOUR OR MORE FORCE VECTORS: THE METHOD OF TRANSPOSITION

When n forces are acting *in the one plane*, these may be combined by using the method of link polygons (*see Figure 2.9*) or the method of calculation (resolved components, *see* Example 2.1 and *Figure 2.11*).

The conditions for *equilibrium* of three forces no longer hold for a system of four or more forces. Whilst it is true that four co-planar forces which intersect at the one point and which formed a closed polygon, are in equilibrium, it is no longer *essential* that they should intersect at the one point. *Figures 2.18(a)* and (*b*) show two examples in which the forces have no common point of intersection, still, as long as the force polygon is closed, both linear and rotational resultants are zero.

47

When the number of forces to be combined is large, it is advisable to choose a *space co-ordinate system*; in this, each force will have components along axes x, y and z. Combination of the forces will then be effected by using *transposition* as

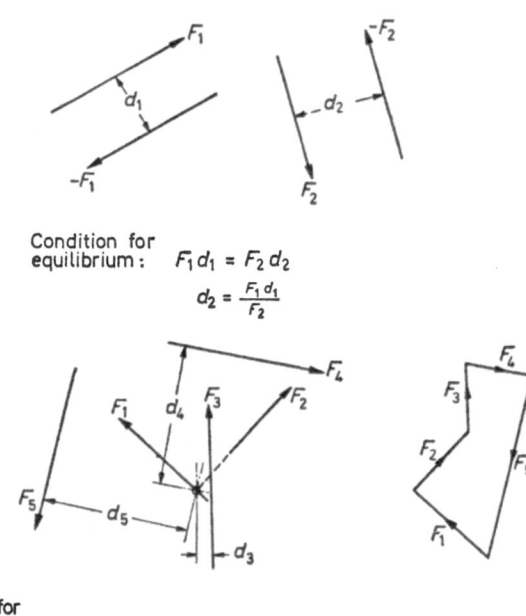

Condition for equilibrium: $F_1 d_1 = F_2 d_2$

$$d_2 = \frac{F_1 d_1}{F_2}$$

Condition for equilibrium: $F_3 d_3 + F_5 d_5 = F_4 d_4$

Force scale

Figure 2.18. Examples of planar equilibrium. (a) Two couples in equilibrium. (b) Five forces in equilibrium

explained in *Figure 2.7*, the *origin of the co-ordinate system* being the point to which forces are transposed.

Figure 2.19(a) shows, by way of illustration, *one* force: F (components F_x, F_y and F_z) acting at point A (co-ordinates x, y and z).

48

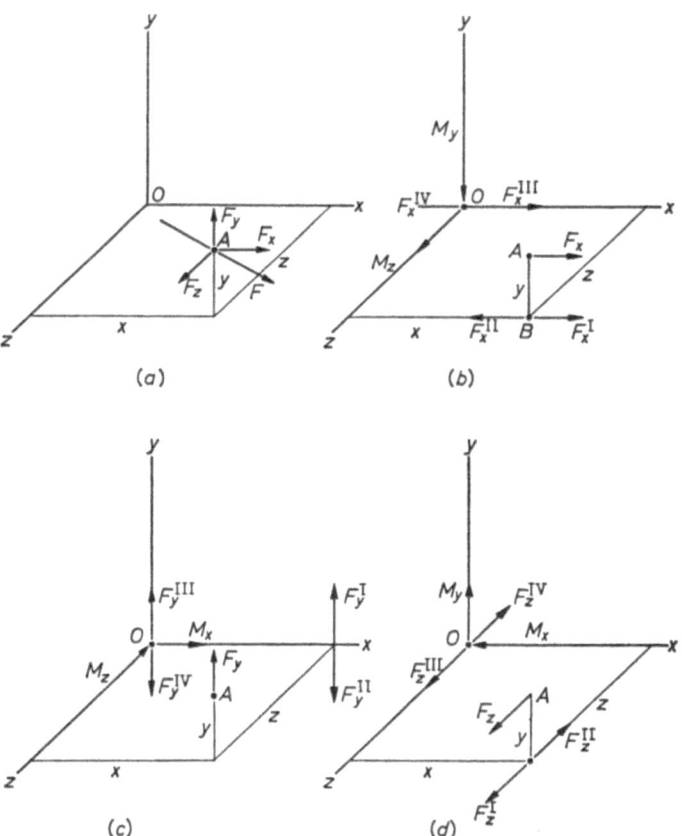

Figure 2.19. Transposition of one force. (a) Force F *acting at point* A. *(b) Transposition of* F$_x$ *to* O. *(c) Transposition of* F$_y$ *to* O. *(d) Transposition of* F$_z$ *to* O

49

In *Figure 2.19(b)*, *component* F_x is transposed to O as follows: Opposite forces $F_x{}^I$ and $F_x{}^{II}$ are introduced at B and opposite forces $F_x{}^{III}$ and $F_x{}^{IV}$ at O, all equal in magnitude to F_x; since these added forces mutually balance out, total value of the system remains unchanged.

Total effect of the five forces F_x, $F_x{}^I$, $F_x{}^{II}$, $F_x{}^{III}$ and $F_x{}^{IV}$ will be:

$$\text{couple } F_x, F_x{}^{II}: \qquad F_x . y = M_z$$
$$\text{couple } F_x{}^I, F_x{}^{IV}: \qquad F_x . z = -M_y$$
$$\text{resultant force}: \qquad F_x{}^{III} = F_x$$

Note that M_z is *positive* since its *rotation is clockwise* when *viewed towards* O, conversely, rotation of M_y is anti-clockwise. *Figures 2.19(c)* and *(d)* show similar *transpositions to O of F_y* and F_z:

F_y		F_z	
couple $F_y, F_y{}^{II}$:	$F_y . z = M_x$	couple $F_z, F_z{}^{II}$:	$F_z . y = -M_x$
couple $F_y{}^I, F_y{}^{IV}$:	$F_y . x = -M_z$	couple $F_z{}^{II}, F_z{}^{IV}$:	$F_z . x = +M_y$
resultant force:	$F^{III} = F_y$	resultant force:	$F_z{}^{III} = F_z$

Summing up:

Force resultant $= F (F_x, F_y, F_z)$ at O.
Moment resultant $= M (M_x, M_y, M_z)$ at O,

where

$$M_x = F_y . z - F_z . y$$
$$M_y = F_z . x - F_x . z$$
$$M_z = F_x . y - F_y . x$$

and $F = \sqrt{F_x{}^2 + F_y{}^2 + F_z{}^2}$; $M = \sqrt{M_x{}^2 + M_y{}^2 + M_z{}^2}$.

The sign of products $F_y . z$ etc. obeys a cyclic order. E.g. in M_x, F_y and z; x, y and z are in *proper sequence* and therefore $F_y . z$ is *positive*. M_y, F_z and x and M_z, F_x and y also follow this sequence. When the reverse sequence occurs, the product has *negative sign*, as in M_x, F_z and y.

When *n forces* act on a body at points A_1, A_2, ... A_n, and these forces are F_1, F_2, ... F_n respectively, then a simple algebraic summation will provide resultants R and M at the origin O of the x, y, z, co-ordinate system:

$$R_x = \Sigma F_x$$
$$R_y = \Sigma F_y$$
$$R_z = \Sigma F_z$$

and $R = \sqrt{R_x{}^2 + R_y{}^2 + R_z{}^2}$ at O.

Further,

$$M_x = \Sigma(F_y . z) - \Sigma(F_z . y)$$
$$M_y = \Sigma(F_z . x) - \Sigma(F_x . z)$$
$$M_z = \Sigma(F_x . y) - \Sigma(F_y . x)$$

and $M = \sqrt{M_x{}^2 + M_y{}^2 + M_z{}^2}$ at O.

Note that the signs of F_x, F_y and F_z must be considered in the summation. These will be positive if the direction of the component coincides with the positive direction of the co-ordinate axis. Naturally, ordinates x, y and z of point A must also be entered with their proper signs. Choice of the co-ordinate system is arbitrary, but once the system and its origin, the point of transposition, is chosen, it must be adhered to throughout the summation.

Example 2.3

The method will be demonstrated on the force system of the cube of *Figure 2.15* which is reproduced as *Figure 2.20*, in which the chosen co-ordinate system is shown with its origin at A.

Products and summations will be carried out in the form of a tabulation in Table 2.1 (page 55). From this,

$$R_x = \Sigma F_x = +1 \text{ ton}$$
$$R_y = \Sigma F_y = +2 \text{ tons}$$
$$R_z = \Sigma F_z = -2 \cdot 5 \text{ tons}$$
$$R = \sqrt{1 + 4 + 6 \cdot 25} = 3 \cdot 35 \text{ tons}$$

51

$$M_x = \Sigma(F_y . z) - \Sigma(F_z . y) = \quad 0 - 10 = -10 \text{ tons-in}$$
$$M_y = \Sigma(F_z . x) - \Sigma(F_x . z) = -5 - \quad 2 = \quad -7 \text{ tons-in}$$
$$M_z = \Sigma(F_x . y) - \Sigma(F_y . x) = -2 + \quad 0 = \quad -2 \text{ tons-in}$$
$$M \quad = \sqrt{4 + 49 + 100} = 12 \cdot 4 \text{ tons-in}$$

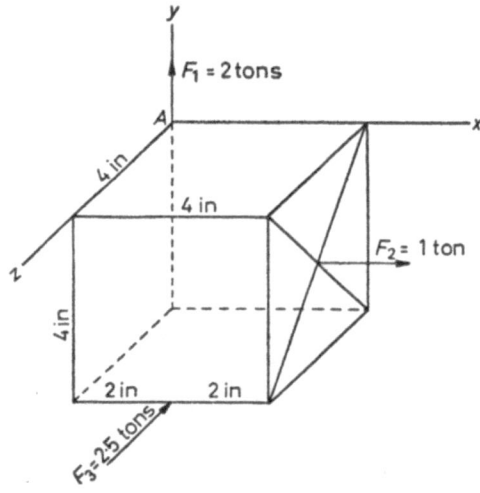

Figure 2.20. Resultant of forces acting on cube

The angles which R or M form with the co-ordinate axes can be readily determined. For example,

$$\text{Angle } R - x = \cos^{-1} \frac{1}{3 \cdot 35} = 73 \text{ degrees}$$

In *Example 2.3*, the essential moves in forming the resultant by transposition were clearly shown. Since the method will be used extensively throughout the book, we will now explain each of these moves in detail:

At the outset, we must have a *concise description* of the force system—magnitude, direction and sense of each of the forces.

A suitable *point for transposition* will be selected and a *co-ordinate system* adopted at the point. Our signs are based on the *x, y, z* system shown in *Figure 2.20*; this will be further discussed in Section 2.7.

F_x, F_y and F_z are the three *components of each force* and these are now determined and entered in the table. When the sense of a component is along the positive co-ordinate axis, the component will be positive. For instance, force F_2 is parallel to the $+x$ axis, so that $F_x = +1$ ton.

Next, the *position ordinates* x, y and z are found for each force. In *Example 2.3*, we used the position ordinates of the point where each force acted on the cube. Again, these ordinates have positive or negative signs. Force F_2, for instance, is acting at the centre of one face of the 4 in cube. The ordinates of this centre point are: $x = +4$, $y = -2$, $z = +2$.

This completes the listing of data in the table and we can now proceed working out *products* and *summations*. There are six products: F_x is multiplied by y and z; F_y by x and z; F_z by x and y. The order of the columns in the table is not important but it is advisable to list the products in the standard manner shown in Table 2.1, for the sake of tidiness. Note that a force component is *not* multiplied by a like position ordinate, for instance, there is no product $F_y.y$.

Totals are formed by *algebraic summation* down the columns of the force components and the products. The *force component* totals are components of the resultant, R. The *product* totals are formed into differences to give components of M as follows:

Each moment component equals the difference between two product totals. For instance,

$$M_y = \Sigma(F_z.x) - \Sigma(F_x.z).$$

That is, for the *y-component* of the moment, the two *product totals made up of x and z quantities* are used. Positive is that product which follows the cyclic sequence in $x - y - z - x - y - z$. In the above, $y - z - x$ is a straight sequence and therefore for M_y, $\Sigma(F_z.x)$ is positive. $y - x - z$ is a reversed sequence, so that in M_y, $\Sigma(F_x.z)$ will be negative.

Once R and M components are known, the *resultant* is practically determined. The sign of R components will show whether they are along the positive or negative co-ordinate directions. The sign of M components indicates if rotations are clockwise (for $+M$ components) if viewed from the positive end of co-ordinate axes.

We may take the solution a step further by forming vector totals of R and similar totals of M components. Directions and angles of these may be obtained from direction cosines.

In conclusion, it should be emphasized that M will be different if another point is chosen for transposition. The student is advised to practise transposition by selecting another point in *Figure 2.20* and repeating the calculation of Table 2.1 for this alternative point.

Example 2.4

Figure 2.21 shows a cantilever steel beam, 8 ft long, embedded in a wall at point O. Resultants should be found with reference to point O and the co-ordinate system chosen there.

Four forces are acting on the beam and in addition, its own weight, at the rate of 56 lb/ft, should also be taken into account; this will be acting at the centre of gravity of the beam, 4 ft from O.

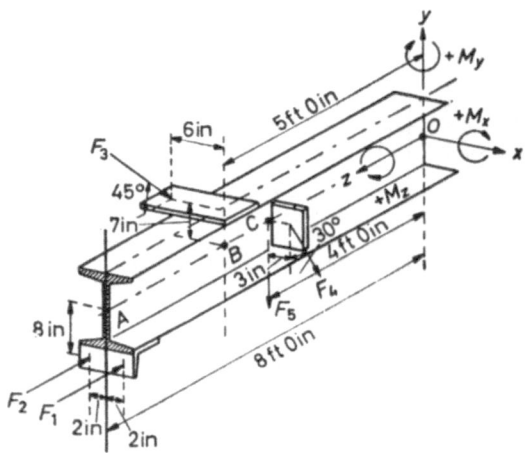

Figure 2.21. Resultant of forces on cantilever beam

List of loads:

No.	Force, tons	Line of force
F_1	2·0	Horizontal
F_2	3·0	Horizontal
F_3	3·0	Parallel to yz
F_4	5·0	Parallel to yx
F_5	0·2	Vertical

Table 2.1. Transposition to A, Example 2.3

Forces	F_x	F_y	F_z	x	y	z	$F_x.y$	$F_x.z$	$F_y.x$	$F_y.z$	$F_z.x$	$F_z.y$
	tons			in			tons-in					
F_1	0	+2	0	0	0	0	0	0	0	0	0	0
F_2	+1	0	0	+4	-2	+2	-2	+2	0	0	0	0
F_3	0	0	-2.5	+2	-4	+4	0	0	0	0	-5	+10
Totals	+1	+2	-2.5				-2	+2	0	0	-5	+10

Table 2.2. Transposition to O, Example 2.4

Forces	F_x	F_y	F_z	x	y	z	$F_x.y$	$F_x.z$	$F_y.x$	$F_y.z$	$F_z.x$	$F_z.y$
	tons			in			tons-in					
F_1	0	0	-2	+2	-8	+96	0	0	0	0	-4.0	+16.0
F_2	0	0	-3	-2	-8	+96	0	0	0	0	+6.0	+24.0
F_3	0	-2.1	-2.1	-6	+7	+60	0	+120	+12.6	-126	+12.6	-14.7
F_4	+2.5	-4.3	0	+3	0	+48	0	0	-12.9	-206	0	0
F_5	0	-0.2	0	0	0	+48	0	0	0	-10	0	0
Totals	+2.5	-6.6	-7.1				0	+120	-0.3	-342	+14.6	+25.3

Table 2.3. Transposition to O, Example 2.5

F_x	F_y	F_z	x	y	z	$F_x.y$	$F_x.z$	$F_y.x$	$F_y.z$	$F_z.x$	$F_z.y$
tons			in			tons-in					
+2.1	-0.7	+1.9	+3	+2	+12	+4.2	+25.2	-2.1	-8.4	+5.7	+3.8

From Table 2.2:

$R_x = +2 \cdot 5$ tons
$R_y = -6 \cdot 6$ tons
$R_z = -7 \cdot 1$ tons $R = \sqrt{2 \cdot 5^2 + 6 \cdot 6^2 + 7 \cdot 1^2} = 10 \cdot 0$ tons

$M_x = -342 - 25 \cdot 3 = -367 \cdot 3$ tons-in
$M_y = +14 \cdot 6 - 120 = -105 \cdot 4$ tons-in
$M_z = 0 - (-0 \cdot 3) = + \quad 0 \cdot 3$ tons-in
$\qquad M = \sqrt{367 \cdot 3^2 + 105 \cdot 4^2 + 0 \cdot 3^2} = 383$ tons-in

This concludes *Example 2.4*.

It may be mentioned that R and M components are numerically equal to the internal force components at *section O* of the beam (ref. Section 1.4).

R_x corresponds to the horizontal shearing force
R_y corresponds to the vertical shearing force
R_z corresponds to the axial thrust force
M_x corresponds to the longitudinal bending moment
M_y corresponds to the lateral bending moment
M_z corresponds to the twisting moment.

We shall return to a further discussion of these internal force components in Chapter 5.

Example 2.5

An attachment to a 15×6 steel I beam is shown in *Figure 2.22* in plan, section and elevation. One force, *F*, is acting on a cleat which is welded to a $5 \times 4 \times \frac{3}{8}$ in angle bracket; this bracket is welded to the web of the beam. Components of *F* are tabulated in Table 2.3. The resultant of the force system $F (F_x, F_y, F_z)$ is to be found with reference to point *O* which is at the centre of the web of the beam, at the section where the vertical centroidal plane *C—y* of the bracket intersects the beam.

$R \quad = \sqrt{2 \cdot 1^2 + 0 \cdot 7^2 + 1 \cdot 9^2} = 2 \cdot 92$ tons $(=F)$
$M_x = F_y \cdot z - F_z \cdot y = -8 \cdot 4 - \quad 3 \cdot 8 = -12 \cdot 2$ tons-in
$M_y = F_z \cdot x - F_x \cdot z = +5 \cdot 7 - 25 \cdot 2 = -19 \cdot 5$ tons-in
$M_z = F_x \cdot y - F_y \cdot x = +4 \cdot 2 + \quad 2 \cdot 1 = + \quad 6 \cdot 3$ tons-in
$M \quad = \sqrt{12 \cdot 2^2 + 19 \cdot 5^2 + 6 \cdot 3^2} = 23 \cdot 9$ tons-in

Further to the remarks made at the conclusion of Example 2.4, it should be noted that results R and M of *Example 2.5* represent the force and moment transferred from the bracket to the beam. R_x ($=F_x$) is the axial thrust acting on the beam, and R_y and R_z are

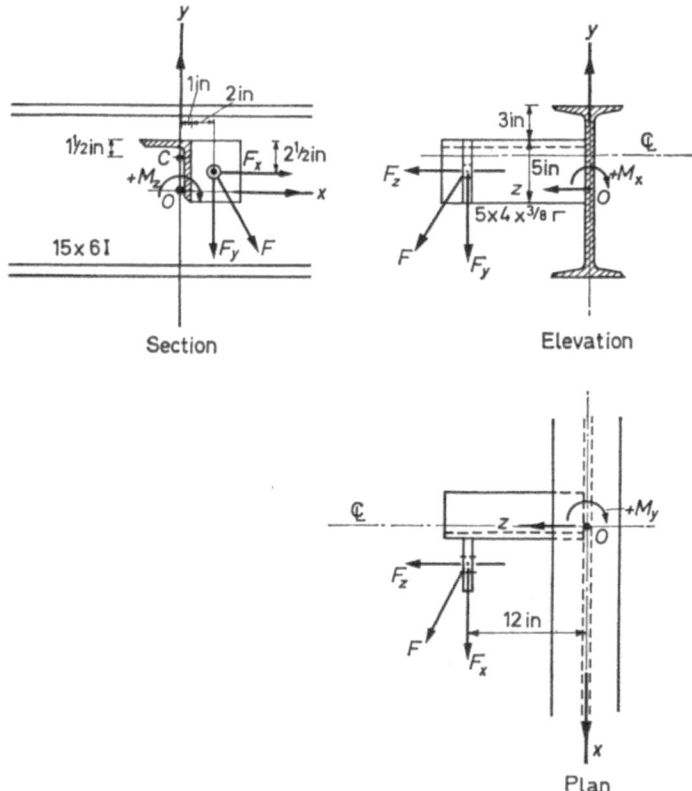

Figure 2.22. Example 2.5

shearing forces. M_x is the twisting moment, M_z and M_y are the longitudinal and lateral bending moments respectively, acting at point O on the beam section. These force components will then be

57

applied to point O in the design of the beam, and will act as loads there.

This completes *Example 2.5*.

2.7. CHOICE OF CO-ORDINATE SYSTEM

The student will note that values of x, y and z can be read direct from a three-projection drawing as soon as the co-ordinate system has been chosen. Choice of the co-ordinate system is arbitrary but care must be taken that the same system is shown on all three projections. In order to use the formulas for M_x, M_y and M_z with the signs as shown above, the co-ordinate system must be the type used when those signs were established, a so-called *right-hand screw* system, as shown again in *Figure 2.23*. It must also be understood that the sign convention for positive M_x, M_y and M_z is that these rotations are positive when they rotate clockwise when viewed towards O.

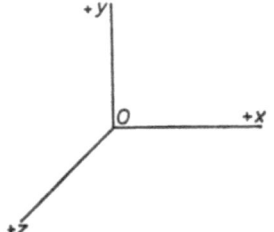

Figure 2.23. Standard (right-hand screw) co-ordinate system

Whatever the co-ordinate system, it is easy to check the sign of $F_x.y$ and similar products. In *Figure 2.24* a *left-hand screw system* is shown, with a positive F_y acting through point A in the x—z plane (pos. x and pos. z). Then $F_y.z$ (positive) will result in $-M_x$ and $F_y.x$ (positive) in $+M_z$. That is,

$$M_x = F_z.y - F_y.z \quad \text{and} \quad M_z = F_y.x - F_x.y:$$

58

signs are opposite to our sign convention with the usual system of co-ordinates.

Figure 2.24. Unconventional (left-hand screw) co-ordinate system

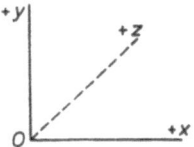

2.8. THE BALANCING FORCE IS OPPOSITE TO THE RESULTANT

In the previous section we have dealt in detail with the combination of forces and the determination of resultants. The magnitude and position of the resultant is of major importance in Dynamics, since the acceleration vector (linear and rotational) is thus determined and the motion of the body is described.

Let us assume now that a force is applied to the body which is equal and opposite to the resultant and its line of action is the same. Under the combined action of the resultant and this opposing force, the body will be in equilibrium. The opposing force is called the *balancing force* and once the resultant of the load system is found, the balancing force is also known, since it opposes the resultant.

In the general case of a load system consisting of n forces,

$$-R \,(-R_x, \, -R_y, \, -R_z)$$

and

$$-M \,(-M_x, \, -M_y, \, -M_z)$$

make up the balancing force system. In order to establish equilibrium, this additional system must be placed on the body.

In this equilibrium condition, the summary of all force and moment components can be listed as follows:

	Resultant	Balancing	
x component of force:	R_x	$-R_x$	$= 0$
y component of force:	R_y	$-R_y$	$= 0$
z component of force:	R_z	$-R_z$	$= 0$

Resultant Balancing

	Resultant	Balancing	
Moment about x axis:	M_x	$-M_x$	$=0$
Moment about y axis:	M_y	$-M_y$	$=0$
Moment about z axis:	M_z	$-M_z$	$=0$

In the special case of *plane equilibrium*, these six conditions reduce to three. Plane equilibrium exists when *all forces* acting (resultant and balancing) are in the one plane.

For example, in the case of a force system which is in the x, y plane:

	Resultant	Balancing	
x component of force:	R_x	$-R_x$	$=0$
y component of force:	R_y	$-R_y$	$=0$
Moment about z axis:	M_z	$-M_z$	$=0$

For the sake of illustration of plane equilibrium, consider the two-engined aircraft shown in *Figure 2.25*. The aircraft is assumed to be in *level forward flight at uniform speed* and the plane of forces will be the x, y plane: in order to bring all forces into this plane, the engine thrusts T will be represented by their resultant, $2T$, which is in this plane.

Thus the acting forces are:

    W the total all-up weight, acting through the C.G., and
    $2T$ the thrust exerted by the two airscrews which are driven
        by the engines.

Aerodynamic effects, induced by the forward motion of the aircraft, provide the balancing forces:

    L_1 the wing lift force
    L_2 the tail lift force, and
    D the total drag (air resistance) for the whole aircraft.

L_1 and L_2 are acting through the *centres of pressure* of the wing and tail surfaces respectively and D acts through the centre of drag. The forces in the x, y plane are shown in the elevation of

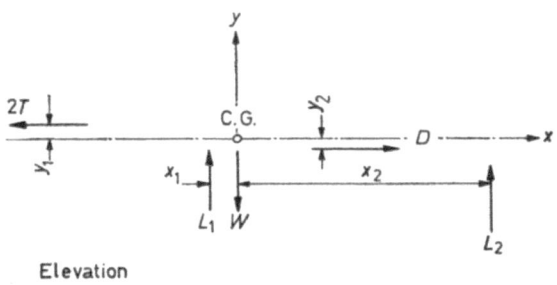

Elevation

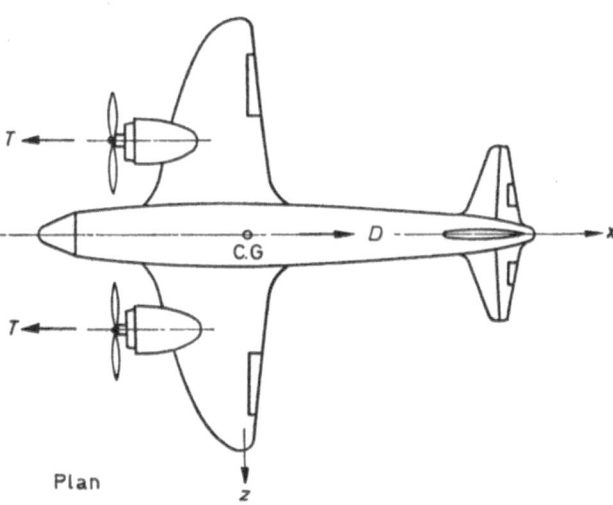

Plan

Figure 2.25. Aircraft in level flight

Figure 2.25. Since this aircraft is in equilibrium in flight, the following must hold:

x component of force: $R_x = 2T - D = 0$
y component of force: $R_y = W - (L_1 + L_2) = 0$

Moment about z axis:

$$M_z = +L_1.x_1 - L_2.x_2 - 2T.y_1 - D.y_2 = 0$$

The moment about the z axis is called the *pitching moment* and if M_z is not equal to zero, the aircraft will be rotated (pitched) about this axis. Atmospheric disturbances cause such pitching, but when the aircraft is inherently stable it will return by itself to the equilibrium position. That is, the rotation will set up resisting (balancing) rotations which return the aircraft to its original attitude.

Similarly, rotation about x (rolling) or about y (yawing) can be caused by such disturbances. Rolling or yawing can be corrected by the pilot who moves the tail elevator or rudder. Permanent corrections for out-of-balance forces can be effected through trim tabs on the wing and on the control surfaces, the adjustment of which provides added aerodynamic forces to balance the aircraft.

2.9. SUMMARY: VECTOR MANIPULATION

Force vectors have both magnitude and direction, and in learning to combine them, we must acquire skill in calculation and in graphical methods.

When combining forces which are in the same plane, a simple graphical construction has advantages. Step-by-step addition using the parallelogram of forces may be used but if there are many forces, the method of link polygon is recommended.

Forces in space are combined using the method of transposition. This process requires the following steps:

(*a*) Choose the point of transposition.

(*b*) Set up a co-ordinate system at the point, *preferably right-hand screw.*

(c) For each acting force, find F_x, F_y, F_z components in this system. When direction of F_x coincides with the positive x axis, F_x will be positive.

(d) For an arbitrary, convenient point on the line of force, find x, y, z ordinates. Again carefully check signs.

(e) In the transposition table, form the six possible products between force components and ordinates: $F_x.y$, $F_x.z$, $F_y.x$, $F_y.z$, $F_z.x$, $F_z.y$. Products are *not* formed between like members, e.g. $F_z.z$ is not calculated.

(f) The algebraic sum of all F_x will give R_x, the x component of the resultant and also for R_y and R_z.

(g) The x-component of the resultant moment vector will be as follows: $M_x = \Sigma(F_y.z) - \Sigma(F_z.y)$.

(h) Here, only the y and z containing products are used for M_x. $F_y.z$ is positive because it follows the proper $x - y - z$ sequence: $M_x - F_y - z$. Conversely, $M_x - F_z - y$ is out of sequence, therefore $F_z.y$ is negative in the formula. The same sequential rule applies to products used for M_y and M_z.

(i) If a right-hand screw co-ordinate system is used, a positive M_x means that it rotates clockwise when viewed from the positive end of the x axis.

(j) The final values of R and M may now be formed as a vector addition from their components.

Once the resultant force and moment are known, an equal and opposite system of force and moment will be the *balancing* force system. If this is added, the combined *total* will be a system in equilibrium.

2.10. PROBLEMS

Problem 2.1

Transpose force system acting on 4 in cube, shown in *Figure 2.26*, to point *O*.

ANS: $R_x = R_y = R_z = 0$,
$\quad M_x = -8$ tons-in, $\quad M_y = +4$ tons-in, $\quad M_z = -4$ tons-in

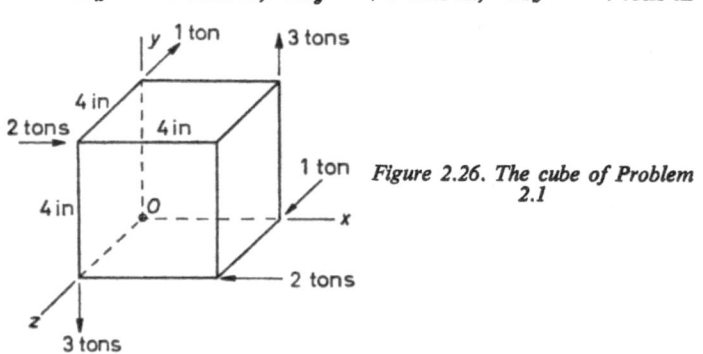

Figure 2.26. The cube of Problem 2.1

Problem 2.2

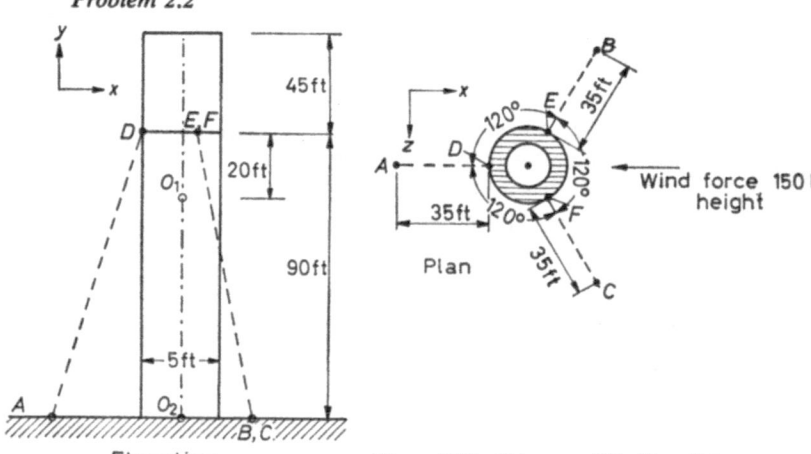

Figure 2.27. Chimney of Problem 2.2

The straight-sided chimney shown in *Figure 2.27* weighs 0·9 ton per ft height. The wind force on the chimney is 150 lb/ft height, acting in the direction shown.

64

The following forces are acting in the guy ropes:

$$\begin{aligned} AD \quad & 0 \\ BE \quad & 6\cdot7 \text{ tons (tension)} \\ CF \quad & 6\cdot7 \text{ tons (tension)} \end{aligned}$$

What is the resultant in (a) Section O_1
(b) Section O_2?

ANS: (a) $R_x = -1\cdot2$ tons, $R_y = -71$ tons, $R_z = 0$
$M_x = M_y = 0$, $M_z = -54\cdot4$ tons-ft
(b) $R_x = -6\cdot6$ tons, $R_y = -134$ tons, $R_z = 0$
$M_x = M_y = 0$, $M_z = -378\cdot4$ tons-ft

Problem 2.3

The cranked beam and loads shown in *Figure 2.28* are in the one plane. Determine resultant.

ANS: $R_x = -9\cdot2$ tons at $y = +4\cdot96$ ft
$R_y = -7\cdot2$ tons at $x = -8\cdot0$ ft $\Big\}$ from G

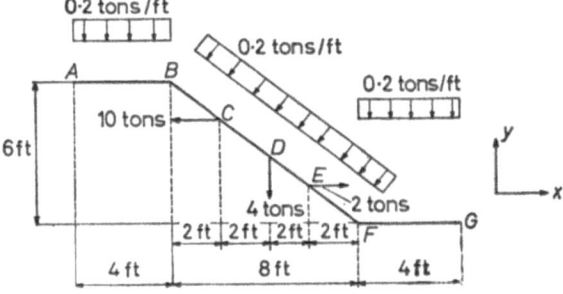

Figure 2.28. Cranked beam of Problem 2.3

Problem 2.4

Figure 2.29 shows section of a continuous retaining wall, the concrete of which weighs 150 lb/ft³. The weight of the fill is 110 lb/ft³ and the pressure of fill and water on face *BC* varies linearly as shown, to a maximum value of 700 lb/ft² at *C*. Is the structure safe against overturning about point *A*?

(Note: It is usual to assume a factor of safety of 1·5 for stability.)

65

ANS: Overturning moment at A = 22,000 lb-ft ⎫ per ft run
Balancing moment at A = 49,500 lb-ft ⎭

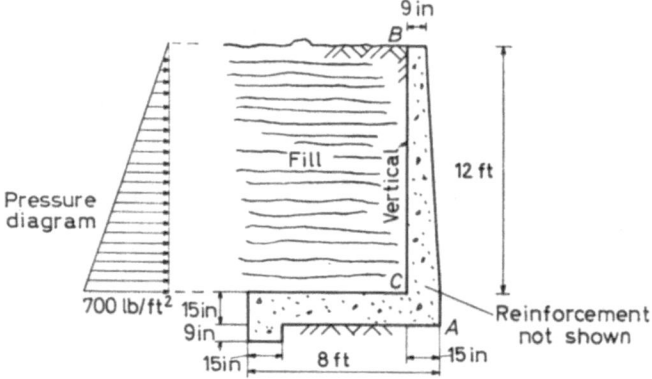

Figure 2.29. Retaining wall of Problem 2.4

Problem 2.5

The cranked rod shown in *Figure 2.30* is acted on by a 2 tons horizontal force at *B*.

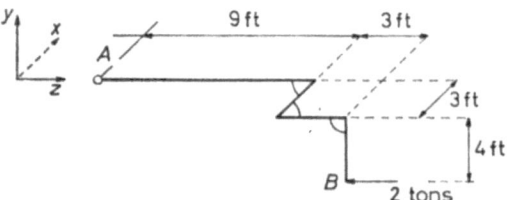

Figure 2.30. Cranked rod of Problem 2.5

What balancing forces are required at *A* in order to hold the rod there firmly?

ANS: $-R_x = -R_y = 0$; $-R_z = +2$ tons
    $-M_x = +8$ tons-ft; $-M_y = -6$ tons-ft; $-M_z = 0$

Problem 2.6

The pendulum shown in *Figure 2.31* is held at an angle of 15 degrees from the vertical by a horizontal force *F*. If the pendulum weighs 5 lb, find the force *F* and the force *T* in the cord.

ANS: $F = 1·34$ lb
$T = 5·18$ lb (tension)

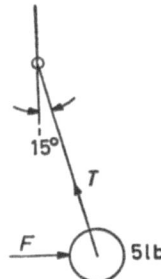

Figure 2.31. Pendulum of Problem 2.6

Problem 2.7

The forces acting on a regular octagonal shape are shown in *Figure 2.32*. Find graphically the magnitude, direction and position of the resultant force.

ANS: $R = 58$ lb, through *A*

Figure 2.32. Force system of Problem 2.7

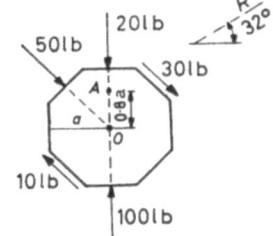

Problem 2.8

The concrete gravity dam shown in *Figure 2.33* has water pressure loads from both sides, in addition to its weight. The dam is 50 ft

67

long. What is the magnitude and direction of the resultant force acting on the foundation? By means of a link polygon find the point of intersection of the resultant force with the base of the dam.

ANS: 3,130 tons at 15 degrees from vertical
14 ft from A

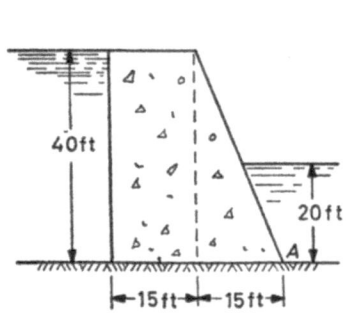

Figure 2.33. Dam of Problem 2.8

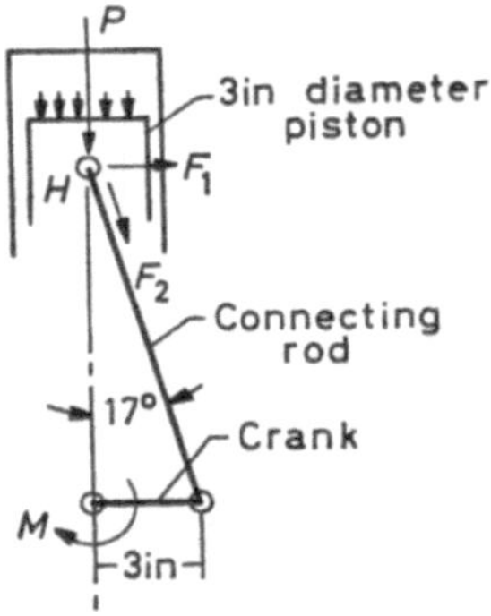

Figure 2.34. Engine detail of Problem 2.9

Problem 2.9

In the internal combustion engine of *Figure 2.34* the force due to the gas pressure in the cylinder is transmitted to the crank by way of the connecting rod. For a gas pressure of 100 lb/in² find the force in the connecting rod and the moment applied to the crankshaft for the position shown. Neglect friction and piston inertia.

Note: Pressure force P is resolved at hinge H into force F_1 and connecting rod force F_2.

ANS: 738 lb (compression)
2,120 lb-in

Problem 2.10

A steel door 7 ft high and 3 ft wide weighs 100 lb and is supported on two hinges 1 ft from the top and bottom of the door. If each hinge takes half of the weight of the door, what is the *resultant* force transmitted to each hinge?

ANS: 58·3 lb at 31 degrees to vertical
(towards the door at the top hinge, towards the frame at the bottom hinge)

68

Problem 2.11

Three force vectors F_1, F_2 and F_3 are shown to scale in *Figure 2.35*. The side of the cube represents 1 ton on the force scale and 2 ft on the length scale. Find force and moment components at A.

ANS: $F_x = +1$ ton, $\quad F_y = -1$ ton, $\quad F_z = 0$
$\quad\quad\ M_x = -2$ tons-ft, $\quad M_y = -2$ tons-ft, $\quad M_z = 0$ ton-ft

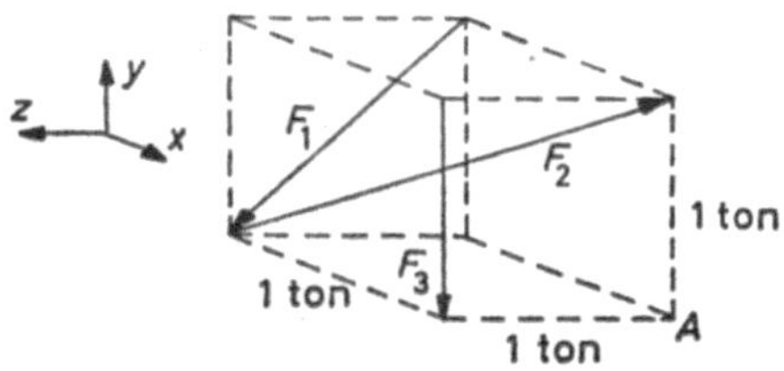

Figure 2.35. Cube of Problem 2.11

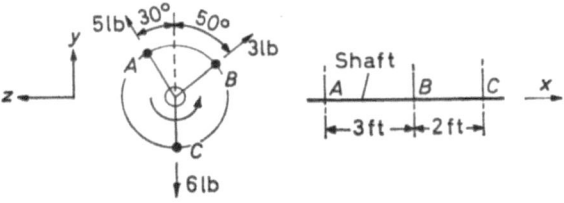

Problem 2.12

Figure 2.36 shows an arrangement of three weights on a rotating shaft in front and side view. The centrifugal forces due to the rotation of the weights are indicated in the diagram. What are the components of the resultant at A?

ANS: $R_x = 0$, $\quad R_y = +0{\cdot}26$ lb, $\quad R_z = +0{\cdot}22$ lb
$\quad\quad\ M_x = 0$, $\quad M_y = -6{\cdot}9$ lb-ft, $\quad M_z = +24{\cdot}2$ lb-ft

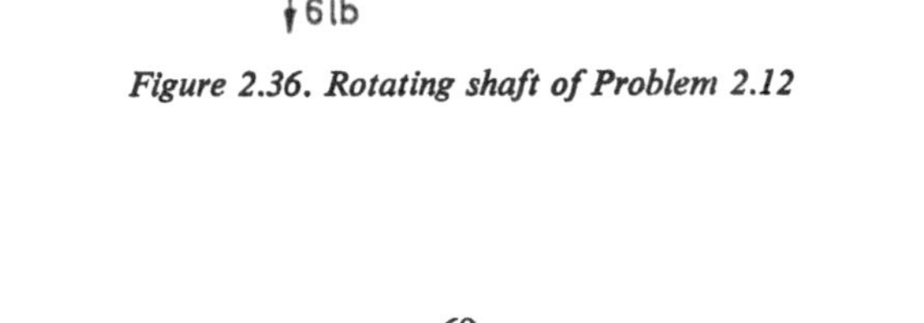

Figure 2.36. Rotating shaft of Problem 2.12

CHAPTER 3

PRINCIPLES OF EQUILIBRIUM

3.1. Effect of supports on internal forces

In the foregoing, methods were introduced which enabled us to form the total force system into a single *resultant*. When this force system is acting on a body, equilibrium is possible only if an *opposite balancing* force is exerted by the body (or its supports). The six components of the acting resultant must be equal, and opposite, to the corresponding six components of the balancing force.

In a different form, the resultant of the acting forces and the balancing force put *together will be zero in all six components.*

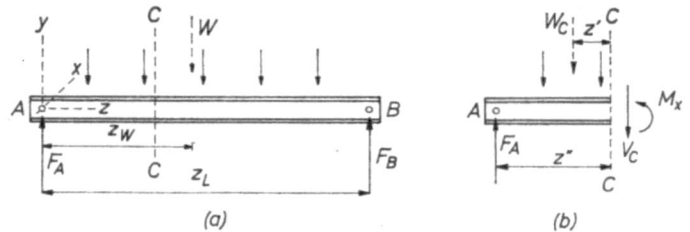

Figure 3.1. Beam reaction

In a given structure or machine part, the acting loads may be specified—such as the weight of the structure, live loads fixed by Building Regulations or the driving loads on a belt or chain. What remains to be determined, however, is the system of support forces or reactions which normally provide the balancing forces. Since the structure is in equilibrium under the effect of *all* forces (acting *and* balancing), the design cannot proceed until the support forces have been found.

Beam AB in *Figure 3.1(a)* is loaded with external forces in the plane of the beam. We assume that these forces have a resultant W. Obviously, the balancing forces F_A and F_B must hold W in equilibrium:

At point A

$$R_y: \qquad F_A + F_B - W = 0 \qquad \qquad \ldots (1)$$
$$M_x: \qquad W.z_W - F_B.z_L = 0 \qquad \qquad \ldots (2)$$

These two equations define F_A and F_B.

From (2): $$F_B = \frac{W.z_W}{z_L}$$

and from (1): $\quad F_A = W - F_B = W\left(1 - \dfrac{z_W}{z_L}\right)$

For a design of this beam, the internal forces in all sections of the beam must be known. Section C—C may be examined by taking a 'cut' through C—C and finding the internal forces at C—C required to hold in equilibrium the acting forces on the portion to the left of C—C.

In Section C—C

This examination of a portion of the structure and its equilibrium under 'acting' and 'internal' forces was discussed in Section 1.4 and is shown in *Figure 3.1(b)*.

The resultant of the loads acting on portion AC (only) is W_C and the internal force required to prevent sliding *along* C—C, the force V_C, is found by forming y-components:

$$R_y = F_A - W_C - V_C = 0$$
that is, $\qquad V_C = F_A - W_C \qquad \qquad \ldots (3)$

V_C is the *balancing force* required in the y-direction along section C—C. It may also be stated that V_C is the *shear resistance* of portion *BC* which prevents *AC* from sliding past it, as explained in Section 1.4.

Design of members must ensure that they are capable of exerting this resistance in *every section*. Thus we are committed to examine the structure thoroughly, determining values of *V* throughout and producing a design, every section of which can safely offer this shearing resistance.

A similar reasoning will lead to internal force components about the other axes. In a general case, six internal force components would be resisted in every section of the body. In the simplified example of *Figure 3.1(a)* and (*b*), these reduce to two: the shearing resistance V_C (*y*-direction) and rotational or bending resistance M_x. Both V_C and M_x depend on the value of F_A, both as regards its *magnitude* and *direction*. Equation (3) clearly shows this; a similar formula for M_x is

$$M_x = W_C.z' - F_A.z''$$

where M_x is the balancing internal force for rotation resistance.

Clearly then, the *type of support* is of importance in the design of structures since this determines the character (magnitude and direction) of the reactions. With different values of F_A, the internal forces in the sections will change and this will vary the design.

3.2. Types of support

Supports can be classified according to the restraint or freedom of the structure at the point of support. The six components of a general resultant may be developed simultaneously at a support, that is, a support may offer rigid resistance in translation (three directions) and rotation (about three axes). On the other hand, the support may offer resistance only in certain directions or about some axes whilst allowing free movement for the body in all other ways.

The ball and socket joint of *Figure 3.2*, firmly embedded in rigid foundations, resists translation in any direction, but the structure which it supports is free to rotate about any axis through the joint.

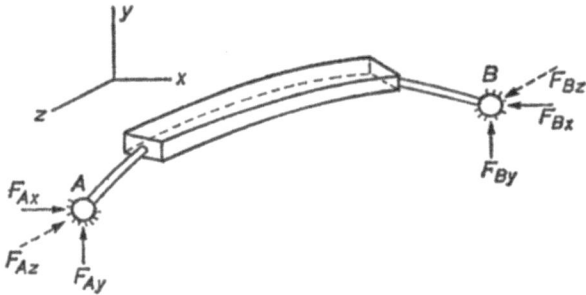

Figure 3.2. Ball and socket joints

The resultant reaction at A is F_A (F_{Ax}, F_{Ay}, F_{Az}) and at B it is F_B (F_{Bx}, F_{By}, F_{Bz}). Resultant moment components of both reactions are zero:

$$M_A (M_{Ax}, M_{Ay}, M_{Az}) = 0$$
$$M_B (M_{Bx}, M_{By}, M_{Bz}) = 0$$

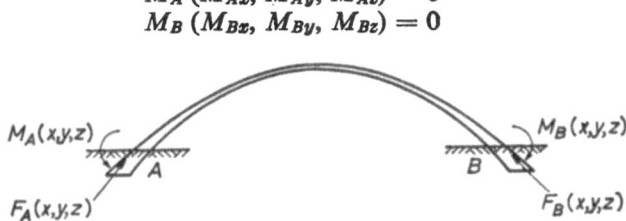

Figure 3.3. Elevation of fully restrained arch, reaction resultant
F *and* M *values*

An extreme case is the fully restrained support such as occurs in the built-in foundations of dams and other major structures. This is shown in *Figure 3.3*.

The term 'fully restrained' is idealized since there is always some slight freedom of movement at the supports. In the arch abutment, the elasticity of the rock will allow movement in all six component directions. This is taken into account in arch design, and thorough exploration of the rock material must be carried out, samples taken and the elasticity of numerous specimens tested in order to ascertain the range of this movement. Since the movement is essentially small, in first approximation the abutments are assumed fully restrained or encased. For calculations of the reactions this presents difficulties since there are *12 components* to be determined (six each at *A* and *B*) and transposing to any point (for instance, to a support) we can only set up *six equations*. Such a structure is called *redundant* or *statically indeterminate* and the reactions are found by applying the rules of elasticity, beyond the scope of this book.

If a space structure has a *total* of *six* unknown *reaction components* (in translation and rotation) then these can be determined for any acting force system. This rule is a mere guide, there are exceptions.

Example 3.1

In *Figure 3.4*, structure *ABC* has one load, *F_D*, acting obliquely. Support *A* is a hinge, i.e. it acts as *A* or *B* in *Figure 3.2*. Support *B*

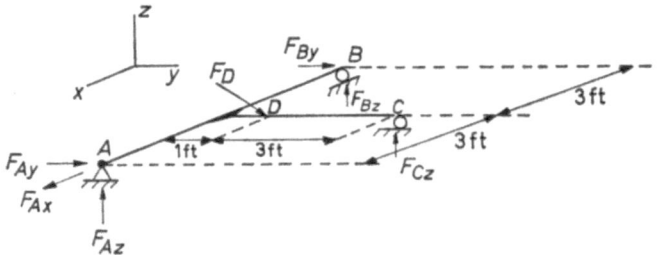

Figure 3.4. The structure of Example 3.1

is a roller which allows complete freedom of rotation but is restrained from moving in the *y* or *z* directions. Support *C* has complete freedom in rotation, and freedom in translation along *x* and *y*.

Table 3.1. Transposition to A

	F_x	F_y	F_z	x	y	z	$F_x.y$	$F_x.z$	$F_y.x$	$F_y.z$	$F_z.x$	$F_z.y$
	tons			ft			tons-ft					
F_D	-5	$+7$	-4	-3	$+1$	0	-5	0	-21	0	$+12$	-4
F_{By}	0	$+F_{By}$	0	-6	0	0	0	0	$-6F_{By}$	0	0	0
F_{Bz}	0	0	$+F_{Bz}$	-6	0	0	0	0	0	0	$-6F_{Bz}$	0
F_{Cz}	0	0	$+F_{Cz}$	-3	$+4$	0	0	0	0	0	$-3F_{Cz}$	$+4F_{Cz}$
Total	-5	$+7$ $+F_{By}$	-4 $+F_{Bz}$ $+F_{Cz}$				-5	0	-21 $-6F_{By}$	0	$+12$ $-6F_{Bz}$ $-3F_{Cz}$	-4 $+4F_{Cz}$
Add at A:	$+F_{Ax}$	$+F_{Ay}$	$+F_{Az}$									

The following reaction components, then, will exist.

At A: $\quad F_{Ax}, \quad F_{Ay}, \quad F_{Az}$
At B: $\qquad\qquad\quad F_{By}, \quad F_{Bz}$ } a total of six components.
At C: $\qquad\qquad\qquad\qquad\quad F_{Cz}$

We may consider now the equilibrium of the body as a whole. If the entire force system (F_D and reactions) is transposed, say, to A, for equilibrium all components of the resultant R and M must be zero. This will yield six equations from which the six desired reaction components can be determined. The choice of point A for transposition is arbitrary and any other convenient point may be used.

Transposition to A of all forces on structure ABC is carried out in Table 3.1.

Equations:

R_x: $-5 + F_{Ax} = 0$ $\qquad\qquad M_x$: $-(-4 + 4F_{Cz}) = 0$
R_y: $+7 + F_{By} + F_{Ay} = 0$ $\qquad M_y$: $12 - 6F_{Bz} - 3F_{Cz} = 0$
R_z: $-4 + F_{Bz} + F_{Cz} + F_{Az} = 0$ $\quad M_z$: $-5 - (-21 - 6F_{By}) = 0$

From these,

$F_{Ax} = +5$ tons
$F_{Ay} = -4{\cdot}33$ tons $\quad F_{By} = -2{\cdot}67$ tons
$F_{Az} = +1{\cdot}5$ tons $\quad\; F_{Bz} = +1{\cdot}5$ tons $\quad F_{Cz} = +1$ ton

The minus sign means that the assumed direction on *Figure 3.4* is reversed; this occurs for F_{Ay} and F_{By}. This concludes Example 3.1.

Once the reaction components are known, internal forces (thrust, shear, bending moment) can be determined *anywhere* on the structure. The process is the same as that demonstrated in *Figure 3.1(b)* and will be shown in detail in Chapter 5.

It is important to note that the notion of equilibrium was first used to determine the unknown reaction (balancing) forces. When these are found and the body as a whole is known to be in equilibrium, we *can* then determine internal forces, again by applying the principles of equilibrium.

Summing up: transposition to *any* point in space will show whether a body is in equilibrium or not. We may use this principle to find reaction components for a body that is *known to be in equilibrium* under the action of acting and support forces. The *nature* and *type* of *supports will determine these components, also the internal forces in the structure.*

3.3. SUPPORTS IN PRACTICE

In Example 3.1 we have assumed three different types of supports. *Figure 3.3* shows still another type. It is useful to list a variety of space support conditions occurring in practice:

Type of support	Practical forms	Reaction components
Universal hinge	Ball and socket joint Radial-thrust bearing	
Universal roller	Spherical castor	
Rigid anchorage	Built-in elements Cantilever ends Encased arch supports	
Flat roller	Wheel Radial bearing Ring or tube support	
Truss hinge	Support or hinged joint in an articulated space frame	

Table 3.2. Practical forms of supports in space

In all these cases, the support was assumed to be at a *point*. In practice, the reaction components are distributed over a surface or several surfaces at the support. Take the case of the motor car or aircraft landing wheel tyre: for the point of the reaction it is usual to assume the centre of the tyre–pavement contact area. For an aircraft in flight, the support forces on the wing and tailplane are due to air pressure (including suction) and we determine *centres of pressure* and assume support forces concentrated at these centres.

3.4. Special cases: plane structures and loads

In the previous section, we dealt in a general manner with a space structure and loading. Many practical structures can be assumed to be of *plane* form: the structure is of a linear character, that is, its span and height are large compared to its thickness, and its *centre line* lies throughout in a central plane of symmetry. In drawing a plane structure, it is assumed that the plane of our paper is this plane of symmetry.

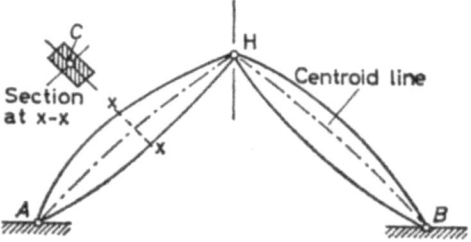

Figure 3.5. Three-hinged plane structure

Figure 3.5 represents a three-hinged plane structure and the dotted centre line shows the position of centroid C throughout the structure.

The equilibrium problem will be simplified if the structure is of *such plane form* and, further, *if all loads act in* its plane. In this case the methods of *graphic statics* can be used to advantage as explained in Chapter 2. Alternatively, *calculations* may still proceed using the transposition principle but they will simplify since the number of support freedoms has been reduced. Since forces are only acting in the plane of the structure, there is scarcely need for support reactions say, at right angles to its plane. The structure will obviously be in equilibrium in that direction since zero *acting* forces cause zero *support* forces.

In *Figure 3.6(a)* a simple beam carries a solitary load which acts in the plane of the beam (i.e. paper).

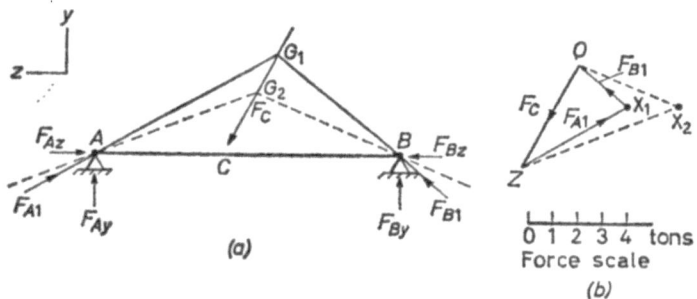

(a)

(b)

Figure 3.6. Redundant plane beam: reactions cannot be found through rules of statics

It can be shown that there is an infinite number of combinations of F_A and F_B, and all satisfy the conditions of equilibrium for the system F_A, F_B, F_C. In the graphical proof, let us plot F_C in a force diagram (*Figure 3.6(b)*). The equilibrium of F_A, F_B, F_C means that the *force diagram* will *be a closed triangle* with sides F_A, F_B and F_C, parallel to the respective force vectors in *Figure 3.6(a)*.

Choose a point on F_C in *Figure 3.6(a)*, say G_1. Since three forces are in equilibrium, this may be the point of intersection of the three forces (*see* Chapter 2). Connect A to G_1, this is a possible line of action for F_A, similarly B to G_1, a likely reaction F_B. In *Figure 3.6(b)* draw parallel lines to these trial F_A and F_B vectors from O and Z respectively. The point X_1 is where these lines intersect and OX_1 and X_1Z give the magnitude of F_B and F_A respectively, in terms of the force scale.

79

Any other point G_2 may be chosen and the same construction repeated. The two trials will supply two different sets of F_A and F_B, both sets satisfying the conditions of equilibrium. This is a typical case of *redundancy* when the conditions of equilibrium are of insufficient number to give a definite solution. It is unlikely that F_{A_1} and F_{B_1} are the *correct* reactions and the 'trial' solutions of *Figure 3.6* of which there can be an indefinitely large number, will *not* yield the beam reactions.

The tools of Statics can only solve the problem if *the number of unknowns* is reduced. The other solution, to produce additional conditions, sufficient for the number of unknowns, is in the realm of Elasticity and does not concern us here.

Certainly, if the *direction* of either F_A or F_B is determined beforehand, the unknowns are reduced and the beam is *determinate*. If B is a *plane roller support*, F_B will be at right angles to the supporting surface as shown in *Figure 3.7(a)*.

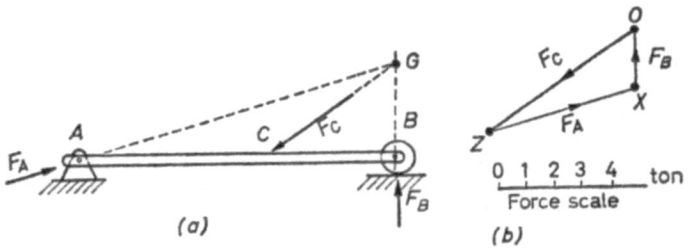

Figure 3.7. Reactions of a determinate plane beam

Point G is found by projecting F_B (along its now known line of action) until it intersects F_C. If we then connect G to A, this will be the only possible line of action of F_A. Similarly, in the force diagram of *Figure 3.7(b)*, we draw a parallel to F_B from O and a parallel to F_A (now of known direction AG) from Z. These two lines will intersect at X and the magnitude of F_A and F_B can be read off the force scale.

Example 3.2

The beam shown in *Figure 3.8(a)* has a hinged support at *A* and a plane roller at *B*. All reaction components are to be determined.

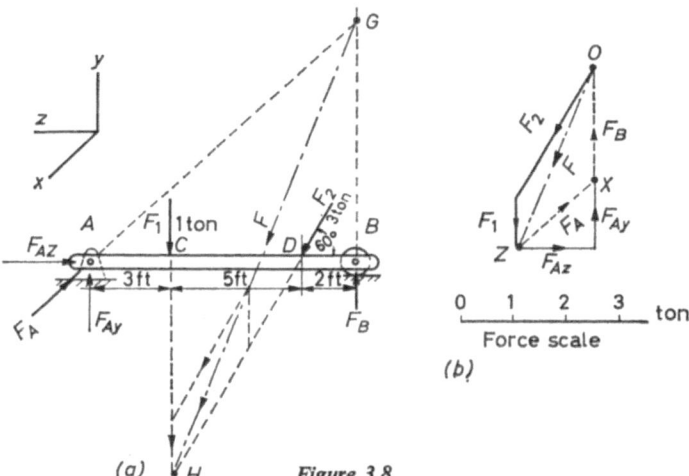

(a)   *Figure 3.8*   *(b)*

Graphical solution

Resultant of loads F_1 and F_2 can be found from *Figure 3.8(a)*, from a parallelogram of forces at *H*, or from the force diagram of *Figure 3.8(b)*. This resultant, *F*, can now be regarded as the only force acting on the beam. Point *G* is at the intersection of *F* and the (known) line of action of F_B, and *AG* will give the line of action of F_A. The problem is completed by reading F_A and F_B from the force diagram, to scale.

Calculation see Table 3.3

$$M_x = F_y . z - F_z . y = 23 \cdot 8 - F_B \times 10 = 0$$
$$F_B = \frac{23 \cdot 8}{10} = + 2 \cdot 38 \text{ tons}$$
$$R_z = -F_{Az} + 1 \cdot 5 = 0$$
$$F_{Az} = +1 \cdot 5 \text{ tons}$$
$$R_y = F_{Ay} + F_B - 3 \cdot 6 = F_{Ay} + 2 \cdot 38 - 3 \cdot 6 = 0$$
$$F_{Ay} = +1 \cdot 22 \text{ tons} \qquad \text{All reactions were as assumed.}$$

Table 3.3. Transposition to A

	F_y	F_z	y	z	$F_y.z$	$F_z.y$
	tons	tons	ft	ft	tons-ft	tons-ft
F_1	-1	0	0	-3	$+3$	0
F_2	$-2\cdot6$	$+1\cdot5$	0	-8	$+20\cdot8$	0
F_A	F_{Ay}	$-F_{Az}$	0	0	0	0
F_B	F_B	0	0	-10	$-F_B\times10$	0
Total	$F_{Ay}-3\cdot6+F_B$	$-F_{Az}+1\cdot5$			$+23\cdot8-F_B\times10$	0

Table 3.4. Transposition to A

	F_y	F_z	y	z	$F_y.z$	$F_z.y$
	lb	lb	ft	ft	lb-ft	lb-ft
F_1	-540	0	0	$+1\cdot5$	-810	0
F_2	$-2,400$	0	$+2\cdot5$	$+7\cdot0$	$-16,800$	0
F_3	-540	0	$+5\cdot0$	$+12\cdot5$	$-6,750$	0
F_4	$-3,780$	$+2,360$	$+2\cdot5$	$+7\cdot0$	$-26,460$	$+5,900$
F_A	$+F_{Ay}$	$-F_{Az}$	0	0	0	0
F_D	$+F_D$	0	$+5\cdot0$	$+14\cdot0$	$+14F_D$	0
Total	$+F_{Ay}+F_D-7,260$	$-F_{Az}+2,360$			$14F_D-50,820$	$+5,900$

Note: When the beam is straight and lies along the z axis, there is no need for the $F_z.y$ column. These values are required though when the beam is bent or cranked.

Example 3.3

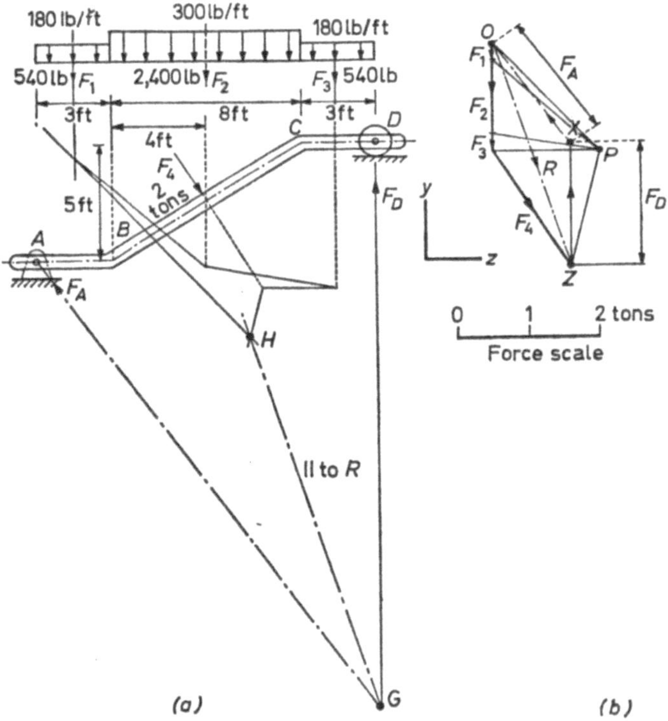

(a)

(b)

Figure 3.9. Example 3.3

Figure 3.9(a) shows the stringer beam of a stair flight (*BC*) with landings at bottom (*AB*) and top (*CD*), both integral with the flight. The support at *D* is assumed a plane roller. The graphical solution is shown in *Figure 3.9(b)*. A link polygon was drawn over the system of

83

forces to determine the resultant of the acting forces, R. The rest of the graphical construction is similar to the method used in *Figure 3.8(b)*. A transposition to A is carried out in Table 3.4.

It is always advisable to assume each reaction component acting along its $+$ co-ordinate direction. Then, if the result is positive, the component direction was *correctly* assumed and it is pointing along the $+$ co-ordinate axis.

1. $F_{Az} = +2,360$ lb (right to left as assumed)
2. $F_{Ay} + F_D - 7,260 = 0$
3. $M_x = 14F_D - 50,820 - 5,900 = 0$

From 3: $F_D = \dfrac{56,720}{14} = 4,050$ lb (up, as assumed)

From 2: $F_{Ay} = 7,260 - 4,050 = 3,210$ lb (up, as assumed).

This completes Example 3.3.

3.5. PLANE SUPPORTS IN PRACTICE

Table 3.5 lists the plane support conditions which may be encountered in practice and the reaction forces which result:

Table 3.5. Practical forms of plane supports

Type of support	Practical forms	Reaction components
Plane hinge	Free structural support	
Plane roller	Sliding structural support Rolling structural support (bridges)	
Rigid encasing (anchorage)	Built-in structural support Welded joint	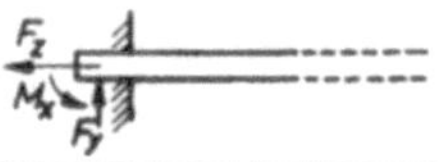
Plane truss hinge	Support or joint of an articulated plane frame (truss)	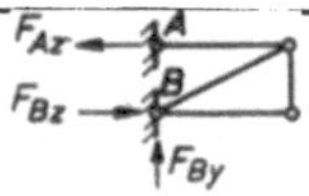

84

In Chapter 4 it will be explained how these types of supports (and the space supports) can be combined to constitute statically determinate systems.

Example 3.4 (*Figure 3.10*)

Cable AC holds up beam BD. B is a plane hinge, A and C are plane truss hinges. Determining the reactions, we note that F_A will be along A_O (for equilibrium of hinge A). The graphical construction first results in R (through H), then the intersection of R and AC gives point G. Since F_B must pass through this point, connecting B and G gives the line of action of F_B.

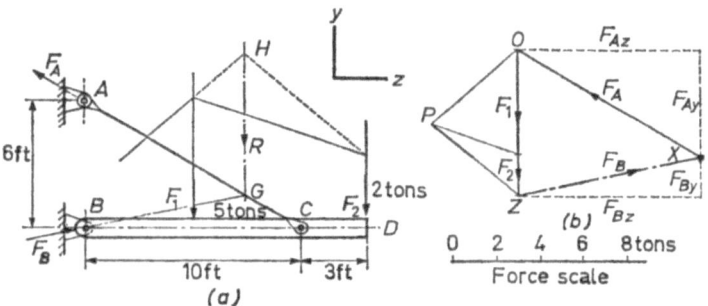

Figure 3.10. Example 3.4

From Table 3.6:

1. $-F_{Az} + F_{Bz} = 0$
2. $+F_{Ay} + F_{By} - 7 = 0$
3. $-6F_{Az} + 51 = 0$ and $F_{Az} = \dfrac{51}{6} = 8\cdot5$ tons (as assumed)

1. $F_{Bz} = F_{Az} = +8\cdot5$ tons (as assumed)   Note that $\dfrac{F_{Ay}}{F_{Az}} = \dfrac{6}{10}$;

   i.e. $F_{Ay} = 0\cdot6F_{Az} = \dfrac{0\cdot6 \times 51}{6} = 5\cdot1$ tons
2. $F_{By} = 7 - 5\cdot1 = 1\cdot9$ tons up (as assumed).

3.6. SUMMARY: PRINCIPLES OF EQUILIBRIUM

A body is in equilibrium under the action of all external forces, that is the sum total of the acting *and* supporting forces. The

Table 3.6. Transposition to B

	F_y	F_z	y	z	$F_y.z$	$F_z.y$
	tons	tons	ft	ft	tons-ft	tons-ft
F_1	-5	0	0	$+5$	-25	0
F_2	-2	0	0	$+13$	-26	0
F_A	$+F_{Ay}$	$-F_{Az}$	$+6$	0	0	$-6F_{Az}$
F_B	$+F_{By}$	$+F_{Bz}$	0	0	0	0
Total	$+F_{Ay} + F_{By} - 7$	$-F_{Az} + F_{Bz}$			-51	$-6F_{Az}$

nature and type of a support will determine the component(s) of the reaction force contributed by the support.

Once the available reaction components are known or assumed, they may be determined by applying the method of transposition. If the structure is *determinate*, the transposition provides *as many equations as there are unknown reaction components*.

In the general case of a *space structure* with acting forces scattered in space, the transposition table provides six equations. This means that a space body which is supported by six reaction components is determinate and these components may be found by solving the six transposition equations.

Dealing with plane structures, the establishment of equilibrium is simplified. If handled graphically, a plane body allows a solution which yields just the sufficient number of reaction components for determinacy of the body. If more components are assumed to exist, then the graphical solution will not have single values but an indeterminately large number of values will be possible. We would have to use the principles of elasticity to solve such cases—this, however, is beyond the scope of this work.

If we use transposition in a plane structure, again we will find that the method results in a number of equations equal to the unknown reaction components if the structure is determinate. As a rule, we can have three unknown components since the plane transposition table supplies three equations.

Once the equilibrium of the *body as a whole* is thus established, the analysis will proceed by determining *internal* force components in selected sections of the body. This will be discussed in greater detail in Chapter 5.

3.7. Problems

Note: + or − signs of answers correspond to + or − co-ordinate direction.

Problem 3.1

Find values of reaction components in the structure of *Figure 3.4* for the following F_D load:

$$F_{Dx} = +6 \text{ tons}$$
$$F_{Dy} = -9 \text{ tons}$$
$$F_{Dz} = -4 \text{ tons}$$

Supports have the following reaction components:

$$\left.\begin{array}{llll} A: & & F_{Ay}, & F_{Az} \\ B: & F_{Bx}, & F_{By}, & F_{Bz} \\ C: & & & F_{Oz} \end{array}\right\} \text{assumed as} + \text{if along the positive axis}$$

ANS: $\quad F_{Ay} = +5.5 \text{ tons}; \quad F_{Az} = +1.5 \text{ tons}$
$F_{Bx} = -6 \text{ tons}; \quad F_{By} = +3.5 \text{ tons}; \quad F_{Bz} = +1.5 \text{ tons}$
$F_{Oz} = +1 \text{ ton}$

Problem 3.2

A motorcar is travelling along a bend in the road which is *super-elevated* as shown in *Figure 3.11*. The car weighs $1\frac{1}{2}$ tons and at its centre of gravity (assumed midway between axles) a 200 lb force is also acting as indicated.

Find reaction components F_{An}, F_{Ap} and F_{Bn} on *one pair* of wheels.

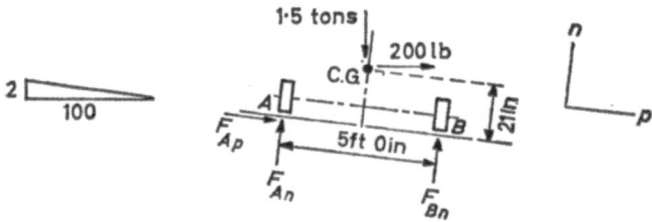

Figure 3.11. Car of Problem 3.2

ANS: $F_{An} = +0.34 \text{ ton}$
$F_{Bn} = +0.41 \text{ ton}$
$F_{Ap} = -100 \text{ lb}$

Problem 3.3

The plane cranked beam *ACDB* shown in *Figure 3.12* carries a 3-ton load at *E*, midway between *C* and *D* and a 2-ton load at *G*, midway between *D* and *B*.

88

Determine the components of reactions at A and B.

ANS: $F_{Ax} = +0\cdot21$ ton
$F_{Ay} = +1\cdot21$ tons
$F_{Bx} = +1\cdot79$ tons
$F_{By} = +1\cdot79$ tons

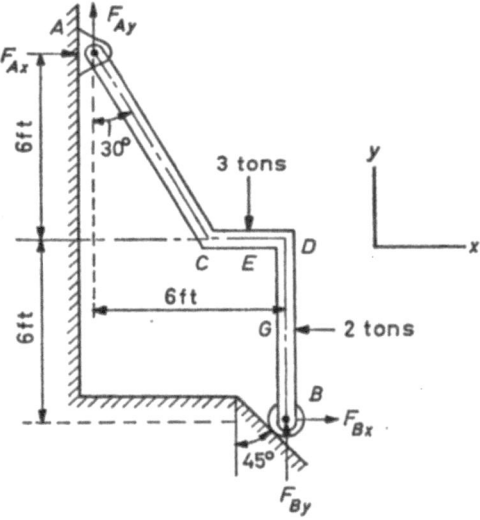

Figure 3.12. Cranked beam of Problem 3.3

Problem 3.4

The tubular chair in *Figure 3.13* is made up of two tubular frames as shown, joined together at A and C with cross tubes. Assuming the loads due to a person's weight and his leaning back, on the whole chair are as indicated, calculate:

(a) The vertical components of support forces at A and B on one frame;
(b) The limiting leaning back force at which the chair's occupier would take off backwards.

The weight of the chair may be neglected.

89

ANS: (a) $F_{Ay} = +68$ lb
$F_{By} = +12$ lb
(b) 51 lb

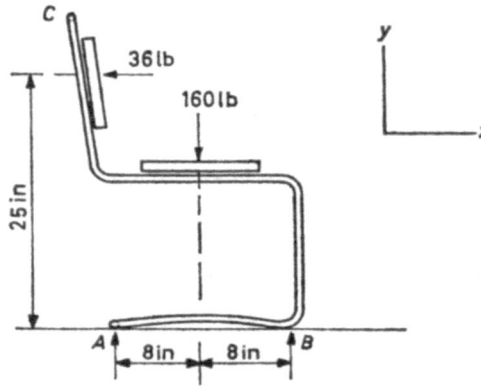

Figure 3.13. Tubular chair of Problem 3.4

Problem 3.5

The plane beam of *Figure 3.14* carries a uniformly distributed

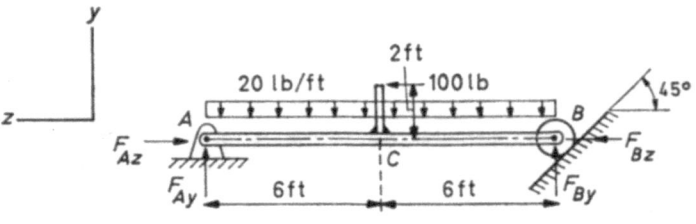

Figure 3.14. Beam of Problem 3.5

load of 20 lb/ft. In addition, a rigidly attached arm at C supports a 100 lb horizontal force at 2 ft from the beam centre line. What are the reaction components?

ANS: $F_{Ay} = +137$ lb
$F_{Az} = -203$ lb
$F_{By} = +103$ lb
$F_{Bz} = +103$ lb

Problem 3.6

A rigid antenna mast is shown in *Figure 3.15*. It is attached to the wall at E and it carries horizontal 100 lb and 50 lb loads and a vertical 200 lb load. Find reaction components at E.

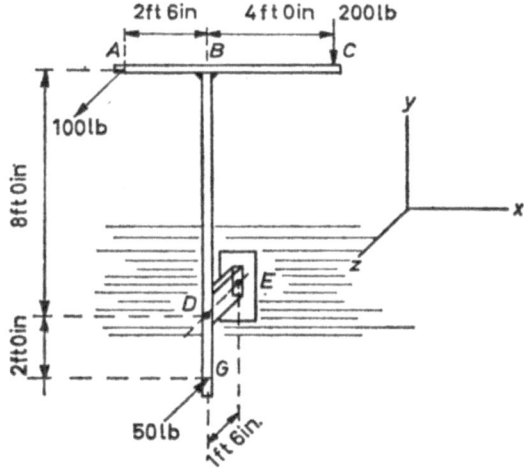

Figure 3.15. Antenna of Problem 3.6

ANS: $F_{Ex} = 0$; $F_{Ey} = +200$ lb; $F_{Ez} = -50$ lb
$M_{Ex} = +1,200$ lb-ft; $M_{Ey} = +250$ lb-ft; $M_{Ez} = -800$ lb-ft

Problem 3.7

Figure 3.16 shows a frame structure, loaded with distributed loading of increasing intensity towards the centre. What are the components of the two support reactions?

ANS: $F_{Ay} = +10$ tons
$F_{By} = +10$ tons
$F_{Bz} = 0$

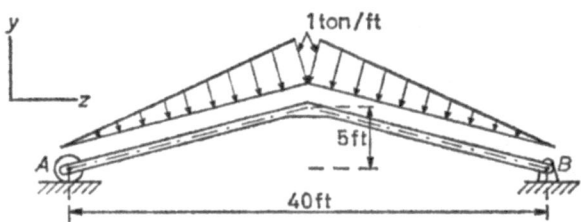

Figure 3.16. Frame of Problem 3.7

Problem 3.8

A straight horizontal shaft, shown in *Figure 3.17*, carries loads which act in vertical planes but at various angles to the shaft. Total weight of the shaft is 50 lb. The shaft is held in tubular supports at

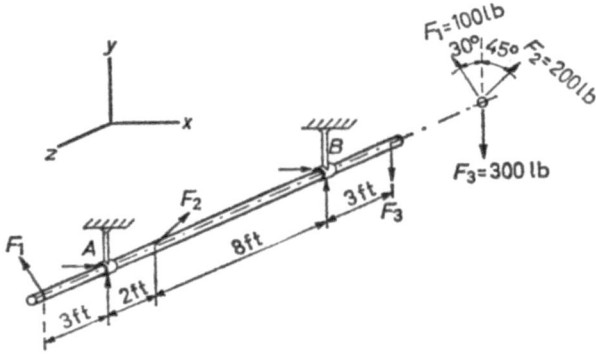

Figure 3.17. Drive shaft of Problem 3.8

A and B which restrain the shaft in both x and y directions. What are the components of the support reactions?

ANS: $F_{Ax} = -47 \cdot 8$ lb; $\quad F_{Ay} = -289 \cdot 6$ lb
$\quad\quad\quad F_{Bx} = -43 \cdot 2$ lb; $\quad F_{By} = +412 \cdot 6$ lb

Problem 3.9

The rigid mast *AC* of *Figure 3.18* is anchored to the ground at *D* with cable *BD*. What are the reaction components at ground supports *C* and *D*?

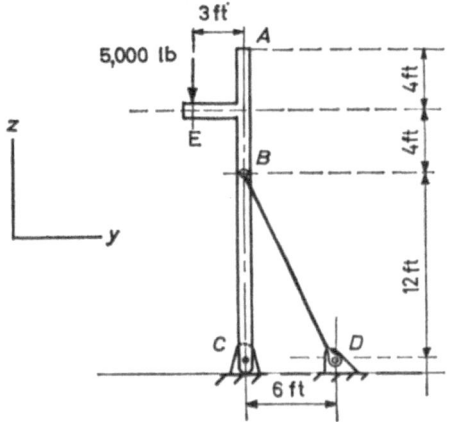

Figure 3.18. Mast of Problem 3.9

ANS: $F_{oy} = -1,240$ lb
$F_{oz} = +7,500$ lb
$F_{Dy} = +1,240$ lb
$F_{Dz} = -2,500$ lb

Problem 3.10

A cantilever beam is shown in *Figure 3.19* with three acting loads. In addition, its own weight of 50 lb/ft is also to be considered. Find reaction components at the encased support *A*.

ANS:
$F_{Ax} = +4$ tons; $\quad F_{Ay} = +2\cdot27$ tons; $F_{Az} = -3$ tons
$M_{Ax} = +26\cdot62$ tons-ft; $M_{Ay} = -22$ tons-ft; $M_{Az} = +2$ tons-ft

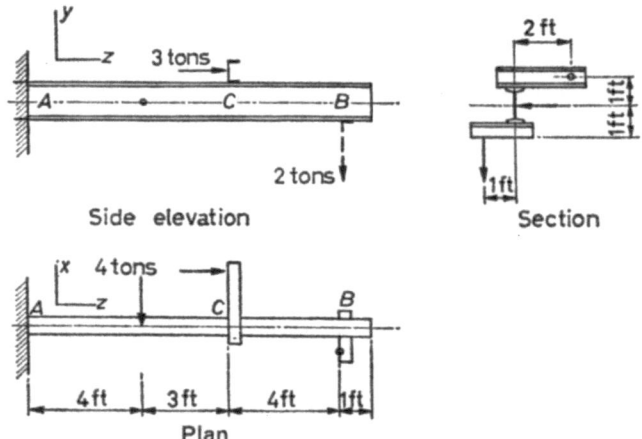

Figure 3.19. Cantilever beam of Problem 3.10

Problem 3.11

The structure shown in *Figure 3.20* has at C a *rigid anchorage*, that is, the frame is a space cantilever from C. Determine reaction components for F_D which has components $F_{Dx} = -1$, $F_{Dy} = -4$, $F_{D\dot{z}} = -2$ tons.

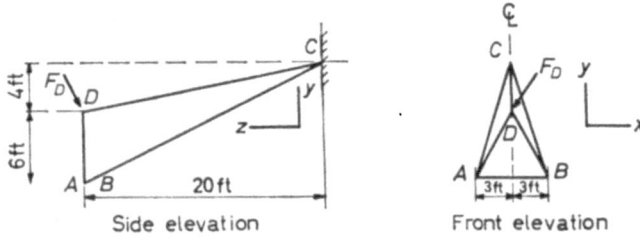

Figure 3.20. Space cantilever of Problem 3.11

ANS:

$F_{Ox} = +1$ ton; $F_{Oy} = +4$ tons; $F_{Oz} = +2$ tons

$M_{Ox} = +88$ tons-ft; $M_{Oy} = -20$ tons-ft; $M_{Oz} = -4$ tons-ft

Problem 3.12

A large illuminated sign contains the letter K shown in *Figure 3.21*, supported at points A, B and C. Neglecting weight of the letter, what are the support reactions for a horizontal force of 100 lb acting at D?

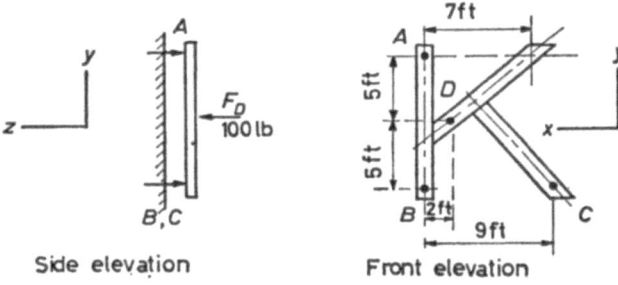

Side elevation   Front elevation

Figure 3.21. Letter K of Problem 3.12

ANS: $F_{Az} = -50$ lb
$F_{Bz} = -28$ lb
$F_{Cz} = -22$ lb

CHAPTER 4

BODIES IN EQUILIBRIUM

4.1. Stability and statical determinacy of a rigid body

As we have seen, a rigid body will be statically determinate if its supports are of just the right kind and of the right number. In this case, the reaction force components can be found by applying the conditions of equilibrium—for instance, by transposing the force system and forming resultants which must total zero.

For a general three-dimensional space body six equations will be obtained, permitting six unknowns to be determined. In the special case of a two-dimensional planar body and force system, three unknown reaction components may be found.

We may now define three major areas of rigid body behaviour under force action:

(a) *unstable bodies (instability);*
(b) *stable and statically determinate bodies;* and
(c) *stable and statically indeterminate bodies.*

Instability should really not be considered here since it leads to motion and this is outside the realm of Statics. It is important, however, to realize the limits of stability and the conditions which lead to instability.

Instability is the result of inadequate support action. It can be transformed into stability by *adding* support restraints. Broadly, the number of restraints (usually reaction components) which are required to change an unstable body into a stable body are its *degrees of freedom.*

In *Figure 4.1(b)* planar beam AB and force F_C are shown. The rigid body, the beam, has three restraints, the three

96

reaction components F_{Az}, F_{Ay} and F_{By}. These can be readily found from the conditions of equilibrium.

(a) Unstable   (b) Stable   (c) Indeterminate
1 degree of   and determinate   1 degree of
freedom   indeterminacy

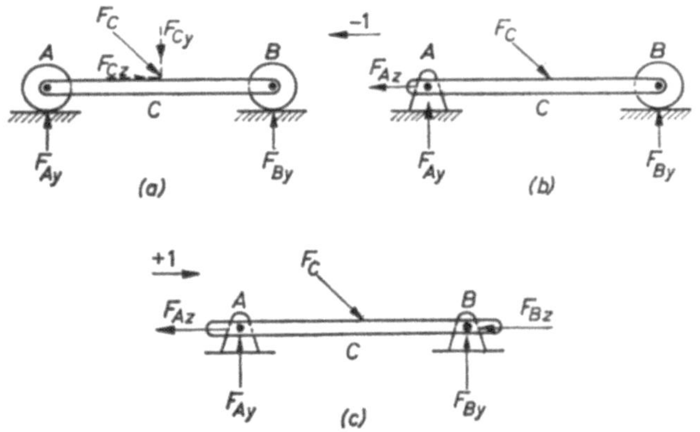

Figure 4.1. Three types of rigid body behaviour

If we take *one restraint* away, say, F_{Az}, the beam in *Figure 4.1(a)* will result. It has *one degree* of freedom, that is, movement along line AB to the right under the action of the horizontal component, F_{Cz} of the acting force F_C. The vertical component F_{Cy} still determines F_{Ay} and F_{By}: these can be readily calculated. There is no way to balance F_{Cz}, however, and the body will move to the right in a straight line motion, with an acceleration $a = \dfrac{F_{Cz}}{m}$ where m is the mass of the beam. This obviously is instability and the one degree of freedom means movement along AB.

We may, on the other hand, have *too many* restraints to a body. In *Figure 4.1(c)* an additional restraint is F_{Bz} which is horizontal, the reaction resulting from the removal of the plane roller at B. Now we have four unknown reaction components, one more than we can evaluate by graphical construction or calculation. The beam of *Figure 4.1(c)* is statically indeterminate and it has *one degree* of statical indeterminacy (one redundancy). Further restraints may be added to increase the degree of indeterminacy, e.g. rotational restraints at A or B.

It is of interest to see the greatest number of degrees of freedom or indeterminacy that can be attained by transforming the beam of *Figure 4.1(b)*. By *removing all three* restraints, the beam will be without any restraints at all, it will be free in space as shown in *Figure 4.2(a)*. This means that the beam has three degrees of freedom: it will translate freely about z and y directions and rotate about x.

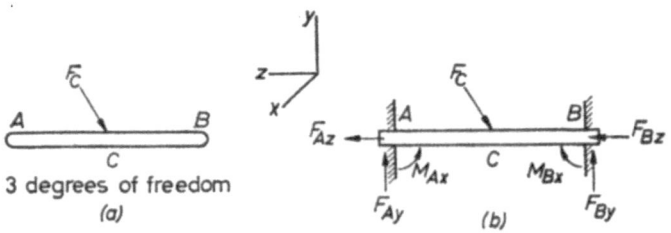

Figure 4.2. Maximum degrees of freedom and indeterminacy for plane beam

The fully encased beam AB of *Figure 4.2(b)* on the other hand, has six support restraints. Thus the maximum degree of indeterminacy is $6 - 3 = 3$.

The example of *Figure 4.1* and *Figure 4.2* is a plane system. Its case applies to any rigid plane body, with a plane force system acting.

In the case of a three-dimensional rigid body, we have already found that six reaction components supply a statically determinate system. Instability will occur by omitting any of these components providing *up to six degrees of freedom* in which extreme case the body is quite unsupported and is free to translate in three directions and rotate about the three axes.

Additional restraints can be introduced at each of, say, three supports. The maximum number of restraints possible is six per support, making a total of eighteen restraints. This means that for this space system we would have a *maximum* of *twelve redundancies*.

Table 4.1 summarizes the frequently occurring conditions explained in the foregoing.

Table 4.1. Summary of conditions of stability and statical determinacy for a rigid body

		Unstable	Stable and statically determinate	Statically indeterminate
		Degrees of freedom	Restraints	Degrees of Redundancy
Plane	1 support	1, 2 or 3	3	0
	2 supports	1, 2 or 3	3	1, 2 or 3
Space	1 support	1 to 6	6	0
	2 supports	1 to 6	6	1 to 6
	3 supports	1 to 6	6	1 to 12

Example 4.1

What is the degree of redundancy of the 'propped cantilever' of *Figure 4.3*?

There are three restraints at A and two at B. Since three restraints make a stable and determinate system, the redundancies are $(3 + 2) - 3 = 2$. An alternative visual method may be employed: without the support at B, the beam would be a cantilever, stable and determinate. That is, we must remove the two restraints at B to make the system stable and determinate, these two restraints represent the redundancies.

This alternative method can be applied also at A: beam AB would be stable and determinate with a plane hinge at B and plane roller at A. That is, F_{Az} and M_{Ax} are redundant components.

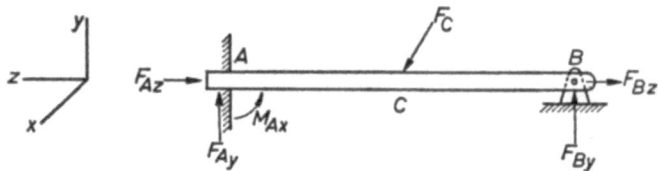

Figure 4.3. Propped cantilever with skew loading

Purely applying conditions of equilibrium, five reaction components cannot be found. We must use rules of elasticity to solve the problem but this is outside the scope of our work.

This completes Example 4.1.

Example 4.2

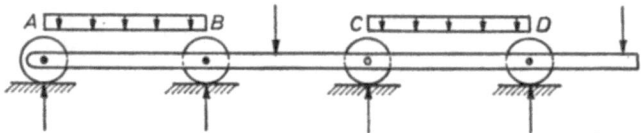

Figure 4.4. Continuous beam

The beam $ABCDE$ of *Figure 4.4* is continuous over four supports. (*a*) Is it stable under *any planar* loading? (*b*) Is it stable and determinate under vertical loading as shown?

(*a*) Under skew loading when the load has a component along the beam, it is unstable since it will move in the horizontal direction. This means an unstable beam with one degree of freedom. If any one of the supports, say A, were to be converted into a plane hinge with two reaction components, the horizontal instability would be taken care of. In that case the beam would be stable and statically indeterminate with two redundancies.

(*b*) Since all loads are vertical, there is no resultant horizontal load and the beam is stable. If we omit the two supports, that is the two (vertical) reaction components at B and C, the remaining beam

100

ADE will be statically determinate since reactions at A and D will be found from conditions of equilibrium. This means that there are two redundancies.

Note that the special vertical loads somewhat change our concepts listed in Table 4.1. On a horizontal plane beam AB which has only vertical loads, we need only *two vertical* reaction components, F_{Ay} and F_{By}. All loads being vertical, there is no need for horizontal restraints for stable equilibrium. Obviously however, as soon as there is a skew load acting, a horizontal reaction component will be required.

4.2. ARTICULATED BODIES

Link supports

In Section 4.1 we assumed supports on solid ground for a rigid body as illustrated in *Figure 4.5*.

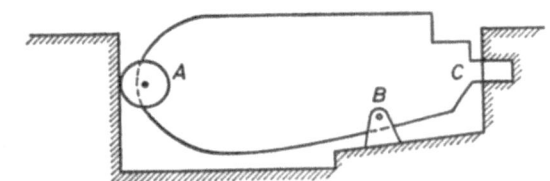

Figure 4.5. Rigid body supported on ground

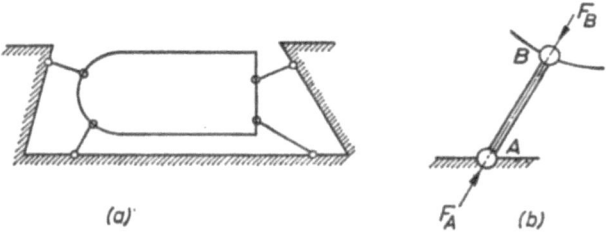

Figure 4.6. Hinged link supports

A rigid body may also be attached to solid supports by hinged links as shown in *Figure 4.6(a)* and (b).

101

Since the hinges at the ends of the links cannot resist rotation, the hinge forces must be along the axis of the link, or, expressed differently, the hinge forces at hinges A and B must be equal and opposite: $F_A = -F_B$.

Example 4.3

The planar system shown in *Figure 4.7* consists of a circular disc, weighing 200 lb and two links, AB and CD, supporting the disc. Determine the support reactions at A and D.

The reactions will be in the direction of the links AB and CD. Graphical construction in *Figure 4.8* is based on the equilibrium of three forces: 200 lb; AB; and CD.

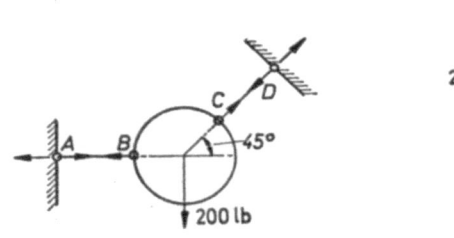

Figure 4.7

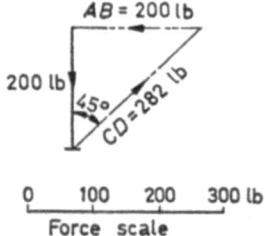

Figure 4.8

Answer: Reaction at A is 200 lb, horizontal, tension
at D is 282 lb, 45 degrees to the horizontal, tension

Note that equilibrium is only possible if the three forces intersect at the one point, in this case at the centre of the disc.

Example 4.4

Figure 4.9 shows a toggle BD which is hinged to solid support at B and linked by AC to the same solid base.

To find the link force AC and the reaction at A, we carry out *graphical construction* in *Figure 4.9(a)* and (*b*).

Results: $AC = F_A = 226$ lb
$F_B = 180$ lb

102

Check by calculation:
Moments about B:
$M_{Bx} = 80 \text{ lb} \times 10 \text{ in} - F_{Ay} \times 5 \text{ in} = 0$
$F_{Ay} = 160 \text{ lb}$
Since $\quad F_{Ay} = F_{Az} = 160 \text{ lb}, \quad F_A = 160 \cdot \sqrt{2} = 226 \text{ lb}$
Note that $F_{By} = F_{Ay} = 160 \text{ lb}$
$F_{Bz} = F_{Az} - 80 = 80 \text{ lb}$
and $\quad F_B = \sqrt{F_{By}{}^2 + F_{Bz}{}^2} = \sqrt{160^2 + 80^2} = 180 \text{ lb}$

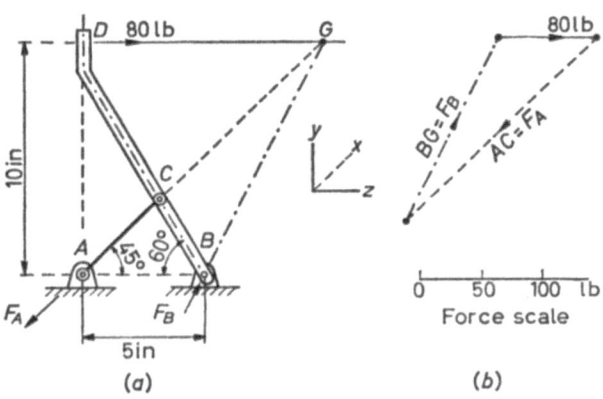

Figure 4.9. Toggle with link

4.3. INSERTED HINGES

The three-hinged frame

Rigid structures may become articulated through the insertion of hinges. This process virtually divides the structure into elements which are separated by hinges. When a structure is statically indeterminate, it is often possible to make it stable and determinate by the insertion of hinges. Generally, one hinge is required to eliminate each degree of redundancy.

The simplest of this type of structure is the 3-hinged frame. In *Figure 4.10(a)*, the rigid body *AB* is supported on plane hinges at *A* and *B*. In this form, it is a stable and statically indeterminate system, with one redundancy. The structure is

103

now severed at C and a hinge is inserted. It is still obviously stable under any planar loading but it also became determinate since all *four* reaction components at A and B can now be found.

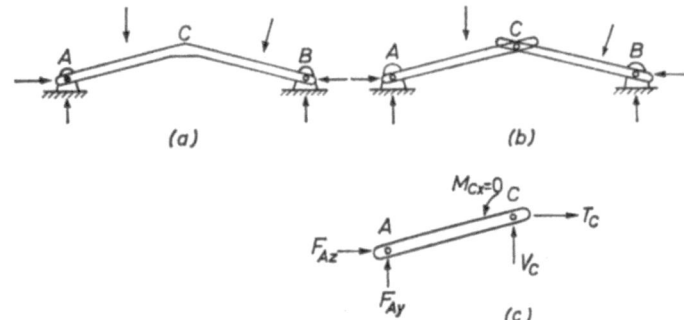

Figure 4.10. Development of a three-hinged frame

In Section 3.11 a relationship was established between reaction and internal forces: in *Figure 3.1(b)* the balancing internal force M_x, in a section of the body AB was calculated considering the equilibrium of portion AC. The reaction at A was one of the forces in the force system on AC.

A similar 'cut' through section (hinge) C in *Figure 4.10(b)* results in portion AC, shown in *Figure 4.10(c)*. Internal forces in *any* cut on a planar body are shear force V_C, axial force T_C and bending moment M_{Cx}. Since C is a hinge, Section C cannot resist rotation and $M_{Cx} = 0$. This supplies an additional condition, which, added to the three conditions of equilibrium, enables us to determine the four reaction components at A and B.

Example 4.5

An industrial, 'saw-tooth' roof is approximated by the outline on *Figure 4.11(a)*. Find all reaction components.

First, missing dimensions will be found (*see Figure 4.11(b)*). From similar triangles ACK and CKB:

$$\frac{16-x}{6} = \frac{6}{x} \text{ that is } x^2 - 16x + 36 = 0 \text{ and } x = 2.7 \text{ ft}$$

and

$$CB = \sqrt{6^2 + 2.7^2} = 6.6 \text{ ft and } AC = \sqrt{16^2 - 6.6^2} = 14.6 \text{ ft}$$

Dimensions and loading are now shown in *Figure 4.11(c)*.

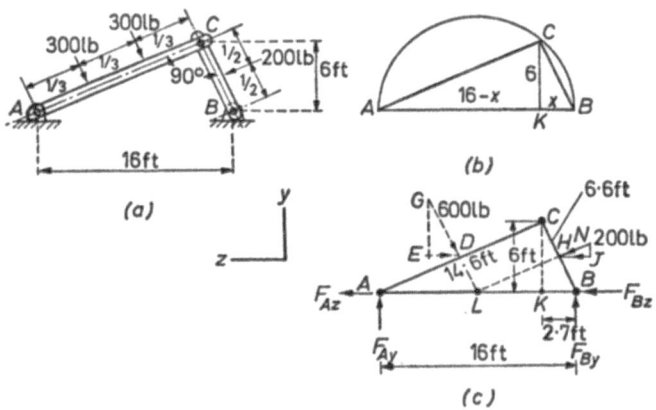

Figure 4.11. Three-hinged saw-tooth roof

Taking moments about B for the entire force system which is in equilibrium:

$$F_{Ay} \times 16 - 600 \text{ lb} \times \frac{14.6}{2} - 200 \text{ lb} \times \frac{6.6}{2} = 0$$

and $\quad F_{Ay} = 315 \text{ lb (up)}$.

Next let us take moments about A:

$$F_{By} \times 16 - 200 \text{ lb} \times \frac{6.6}{2} - 600 \text{ lb} \times \frac{14.6}{2} = 0$$

and $\quad F_{By} = 315 \text{ lb (up)}$.

Note that $F_{Ay} = F_{By}$, even though individual loads are not symmetrical. This is due to the fact that the *overall resultant* of the acting loads happens to pass through L, the *midpoint* of the line AB.

No further equilibrium equations will enable us to calculate F_{Az} or F_{Bz}. However, the fact that the *bending moment at* C must be zero can be written for *the portion* AC:

$$BM_O = F_{Az} \times 6 - F_{Ay}(16 - 2 \cdot 7) + 600 \times \frac{14 \cdot 6}{2} = 0$$

and using the value for F_{Ay} found above:

$$F_{Az} = \frac{315 \times 13 \cdot 3 - 600 \times 7 \cdot 3}{6} = 31 \text{ lb (i.e. right to left)}$$

Similarly for *the portion* BC:

$$BM_O = F_{Bz} \times 6 - F_{By} \times 2 \cdot 7 + 200 \times \frac{6 \cdot 6}{2} = 0$$

that is,

$$F_{Bz} = \frac{315 \times 2 \cdot 7 - 200 \times 3 \cdot 3}{6} = 32 \text{ lb (i.e. right to left)}.$$

We may now *check* on, say, the *horizontal equilibrium* of the structure. Horizontal component of 600 lb load (two 300 lb loads) can be arrived at by using similar triangles GED and CKB: in *Figure 4.11(c)*:

$$\frac{600}{ED} = \frac{6 \cdot 6}{2 \cdot 7}, \text{ that is } ED = \frac{2 \cdot 7 \times 600}{6 \cdot 6} = 245 \text{ lb (left to right)}.$$

To find the horizontal component of the 200 lb load, we use the similar triangles NJH and CKB:

$$\frac{200}{NJ} = \frac{6 \cdot 6}{6} \text{ and } NJ = \frac{6}{6 \cdot 6} \times 200 = 182 \text{ lb (right to left)}.$$

Horizontal equilibrium requires that

$F_{Az} + F_{Bz} + ED + NJ = 0$, and taking right to left as positive, $31 + 32 + 182 - 245 = 0$. This is correct.

Similar checking may be carried out for vertical components.

Results are:

$$F_{Az} = 31 \text{ lb right to left}$$
$$F_{Ay} = 315 \text{ lb up}$$
$$F_{Bz} = 32 \text{ lb right to left}$$
$$F_{By} = 315 \text{ lb up.}$$

This completes Example 4.5.

Note that the insertion of a hinge in a *straight beam* (as shown in *Figure 4.12*) will not result in a stable structure.

Figure 4.12. Unstable three-hinged beam

A small amount of movement must take place at hinge C before the frame becomes stable. This movement may occur up or down, according to the loading. Since such movement presupposes elastic behaviour, this case is outside our scope.

The Gerber-type beam

Insertion of hinges into straight beams is still feasible (and can result in stable structures) if the beam has at least two redundancies. Typical examples are shown in *Figure 4.13*.

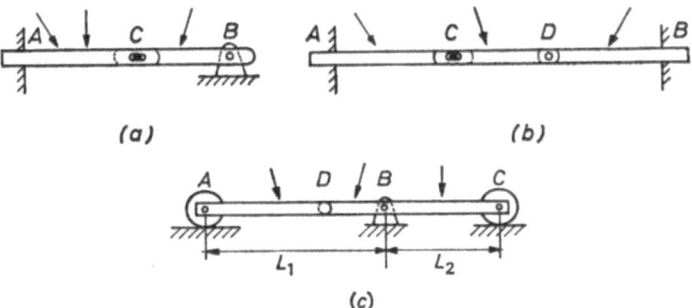

(a)

(b)

(c)

Figure 4.13. Beams made stable and determinate by articulation (a) Propped cantilever. (b) Encased beam. (c) Continuous beam

As before, we have deliberately fixed the *bending moment* at zero at the hinge section. It is as if the hinge acted as a kind of '*internal support*' and, in fact, *Figure 4.13(a)* can be re-drawn

107

as shown in *Figure 4.14(a)*. The hinge at *C can transfer shear* and if we regard *C* as a roller support, the 'beam' *CB* will have the reactions shown at *C* and *B*.

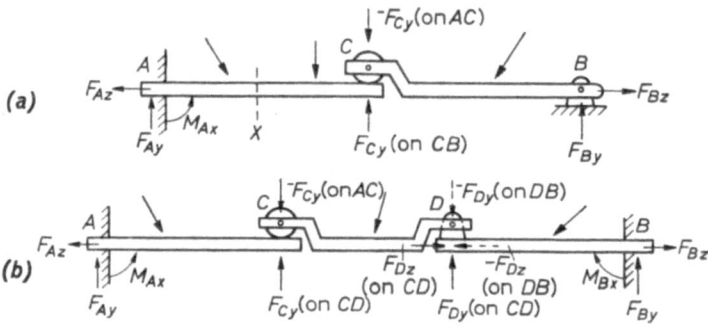

Figure 4.14. Hinges replaced by 'internal supports'

The support reaction F_{Cy} will then be provided by the cantilever *AC*. That is, the load $-F_{Cy}$ will act on the cantilever at *C*, in addition to other loads directly acting on the cantilever. A solution will proceed as follows:

(a) Regard *CB* as a simply supported span.
(b) Calculate reactions (F_{Bz}, F_{By}, F_{Cy}).
(c) Internal forces on span *CB* may now be obtained for any section.
(d) Apply end load $-F_{Cy}$ on cantilever *AC*.
(e) With loads acting on *AC*, this will enable determination of internal forces anywhere on cantilever *AC*.

Note that reaction components at *A* are not required for this since internal forces at section X, say, may be found by considering portion *XC* and all acting forces on this portion are known.

Similar consideration applies to the fully encased beam shown in *Figures 4.13(b)* and *4.14(b)*. Reactions from the 'suspended' span CD are applied to cantilevers and thus the solution is determinate and internal forces may be found.

The beam of *Figure 4.13(c)* is the prototype of the 'Gerber' beam. In this, the insertion of a hinge means *reduction in the maximum bending moment* as well as statical determinacy. (*See Figure 4.15.*)

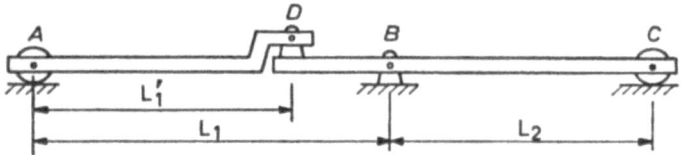

Figure 4.15. Smallest Gerber beam

This occurs for two reasons: L_1 is shortened into an L_1' (suspended) span and L_2 has a cantilever DB which tends to reduce bending moments within span BC. Any number of spans can be produced in this manner and made determinate by the insertion of one hinge in each span as shown in *Figure 4.16(a)*. The solution is shown in *Figure 4.16(b)*.

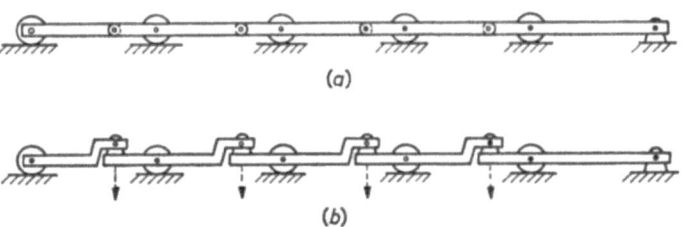

(a)

(b)

Figure 4.16. Multi-span Gerber beam

Some hinges represent rollers, others *plane* hinges as required for stable and statically determinate structures. If vertical loads only are acting on horizontal beams, hinges may as well be just plain hinges as shown in *Figure 4.17(a)*. But if skew loads are acting which have a component *along* the beam, then some hinges must be of the form shown in *Figure 4.17(b)* in order to produce a statically determinate structure—compare with hinge *C* in *Figures 4.14(a)* and *(b)*.

(a)   (b)

Figure 4.17. (a) Plain hinge. (b) Roller (sliding) hinge

Example 4.6

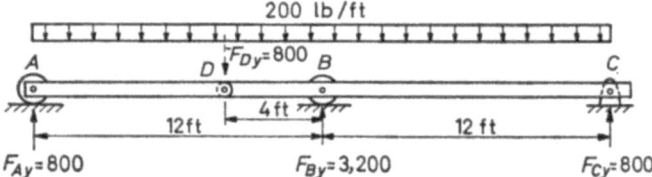

Figure 4.18. Two-span Gerber-type beam

The beam *ABC* is made determinate by the insertion of a hinge at *D*—see *Figure 4.18*. What are the reactions at *A*, *B* and *C*?

Span AD
$$F_{Dy} = \frac{8 \times 200}{2} = 800 \text{ lb}$$
$$F_{Ay} = 800 \text{ lb}$$

Beam DBC

This beam is loaded with 200 lb/ft and the $-F_{Dy} = 800$ lb cantilever load.

Moments about *C*:

$$800 \times 16 + 3{,}200 \times 8 - F_{By} \times 12 = 0$$
$$F_{By} = \frac{12{,}800 + 25{,}600}{12} = \frac{38{,}400}{12} = 3{,}200 \text{ lb}$$

Moments about B:

$$800 \times 4 - 3,200 \times 4 + F_{Cy} \times 12 = 0$$
$$F_{Cy} = \frac{12,800 - 3,200}{12} = \frac{9,600}{12} = 800 \text{ lb}$$

Check: Downward loads are $200 \times 24 = 4,800$ lb
Upward loads are $800 + 3,200 + 800 = 4,800$ lb. This is correct.

Note that in this arrangement the support at B receives a large share of the total load. This share depends on the position of D in span AB. When it is right on B, AB and BC are separate simple beams and $F_{By} = 2,400$ lb. When it has moved to A, beam ABC becomes a two-span indeterminate structure and B would have a reaction of $1{\cdot}25 \times 12$ ft $\times 200 = 3,000$ lb. (The factor of $1{\cdot}25$ is the result of elastic considerations of the statically indeterminate structure.)

4.4. FULLY ARTICULATED STRUCTURES

These consist of straight members which, at their ends, are connected by hinges. As long as *loads are only applied to the end joints* (hinges) and not to the members themselves, internal forces are merely axial thrust forces (tension or compression).

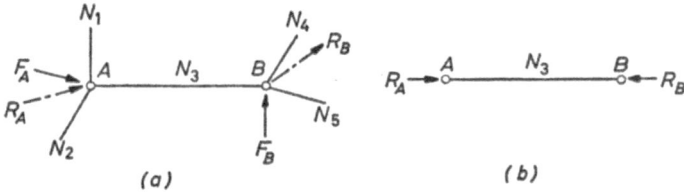

Figure 4.19. Member in articulated structure. (a) *Not possible.* (b) *The only possible arrangement of* R_A *and* R_B

To prove this, let us consider the member AB in *Figure 4.19(a)*. At its hinged end A, members N_1 and N_2 are joined to it and the external force F_A is acting: the *total combined effect* of N_1, N_2

111

and F_A is R_A. Similarly, the resultant at B of N_4, N_5 and F_B is R_B. Obviously, member AB is acted on by *two resultant* forces: R_A and R_B. For AB to be in equilibrium, R_A and R_B must be in equilibrium—but this is only possible if they are equal and opposite, and if they have a common line of action as shown in *Figure 4.19(b)*. Obviously, too, they will act along AB which thus has no internal force component at right angles to its length, hence no shearing force or bending moment.

4.5. STABILITY OF PLANE TRUSSES

Articulated structures are called 'trusses'. The smallest possible plane truss has three members and three joints as shown in *Figure 4.20(a)*.

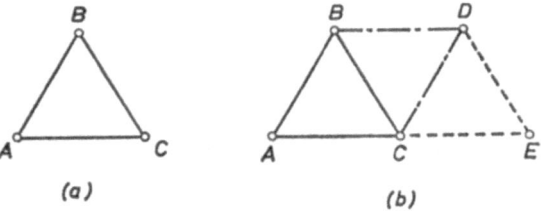

Figure 4.20. Basic plane truss and the building up of a plane truss form

Hinges may be added to this basic truss and these hinges would then be connected to it by using further members; for example hinge D is added, and connected to ABC by new members BD and DC, *see Figure 4.20(b)*. Or, a further hinge E is added by using DE and CE.

In all stages of the adding process, the truss was 'just stiff', that is *stable* and *statically determinate*. If a force system in equilibrium is acting, say, on truss $ABCD$, every internal thrust force in the truss can be calculated and the truss itself would be in equilibrium.

If our truss has *j* number of *joints* (hinges), the total number of *members q* to make it just stiff would be

$$q = 2j - 3$$

since for each joint two members are required but the initial frame of three members is already there. If our truss has fewer members than this, it is unstable. If it has more, it is statically indeterminate, that is, there are too many forces to find from the insufficient conditions of equilibrium.

Example 4.7

The hexagonal frame of *Figure 4.21(a)* represents the bracing structure of a mine shaft. Is it just stiff?

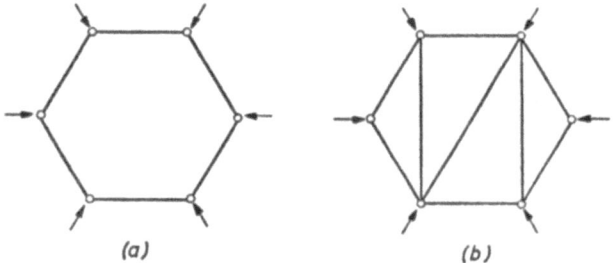

(a)   (b)

Figure 4.21. Frame of Example 4.7. (a) Unstable. (b) Stable and determinate

$q = 2 \times 6 - 3 = 9$ members are required. But we have only 6: the frame as it is shown in *Figure 4.21(a)* is unstable. *Figure 4.21(b)* shows one possible solution for the bracing which makes it stable and determinate.

Naturally, it is assumed that the six external forces form a system in equilibrium. It must follow, then, that the internal forces in the members can be found—these being axial thrust forces.

The above formula applies to trusses which are completely free and it does not account for solid supports. When the truss has j_s hinges (out of a total of *j*) supported on solid ground, these j_s joints need not be supported amongst themselves. That is, to make the truss just stiff

$$q = 2j - 3 - (2j_s - 3) = 2(j - j_s)$$

113

To make the truss just stiff, the number of members will be
twice the free (unsupported) joints.

Example 4.8

Is the truss shown in *Figure 4.22* just stiff? A, B and C are hinges
solidly supported, D and E are free.

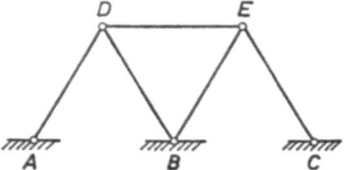

*Figure 4.22. Plane truss of Example
4.8*

If we use the formula:

$$q = 2 \text{ (number of free joints)} = 2 \times 2 = 4$$

Since there are five members, this truss is not just stiff, it has one
degree of redundancy.

In order to make it just stiff, we could remove one member, *any
one* in fact. *Figure 4.23(a)*, (*b*) and (*c*) show three alternatives, all
have four members, all are just stiff.

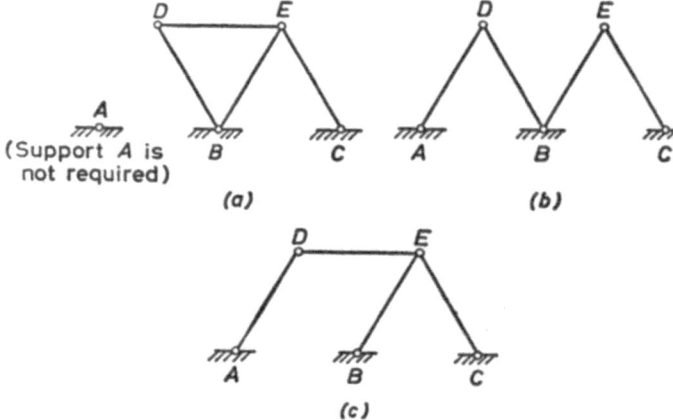

Figure 4.23. Plane truss of Figure 4.22 with a member removed

114

Example 4.9

Figure 4.24 shows an 'open web joist' which is used as a girder to span the distance *AB*. Is it just stiff?

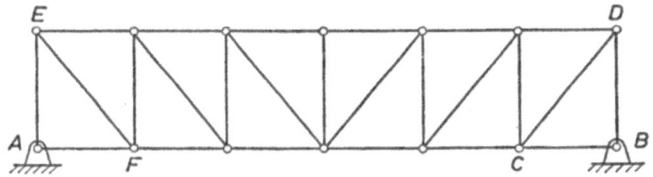

Figure 4.24. Open web joist of Example 4.9

There are 12 free joints and there should be $2 \times 12 = 24$ members. However, our count gives 25, that is, one member is superfluous.

In this case we cannot remove just any of the members. Removal of *DB*, for instance, would imply that the reaction at *B* is horizontal, since, for equilibrium at *B*, the only remaining member *CB* and the reaction R_B should be equal and opposite. A horizontal reaction R_B is unlikely and this arrangement, therefore, is not practical.

On the other hand, removal of *CB* looks logical since this would break the direct link between *A* and *B*, which is superfluous since both are on solid ground.

Thus the reaction at *B* would be defined as being equal and opposite to the force in member *BD*. This is really equivalent to placing a plane roller under *B* which is in any case required for making the structure statically determinate: if we regard the girder as a solid beam *AEDB*, the reactions at *A* and *B* would remain the same. We have seen earlier, however, that these reactions are determinate only as long as *the direction* of one reaction is defined. By defining R_B as being along member *BD* due to removal of member *CB*, the girder (truss or beam) will be determinate.

When only vertical loads act on the girder, both R_A and R_B are vertical and R_A will balance *AE*, R_B will balance *BD*. Then forces in both *AF* and *CB* are zero and both these members would theoretically be superfluous for a stable structure. The reason for placing them into the girder is that *A* and *B* may move and the structure may accidentally receive horizontal loads. Then reactions at *A* and *B* will no longer be vertical.

115

If there is an actual plane roller under *B*, the member *CB* is required to avoid excessive movement of the roller under accidental lateral loading.

Example 4.10

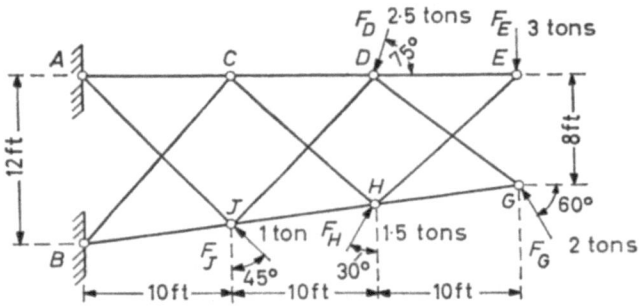

Figure 4.25. Cantilever truss of Example 4.10

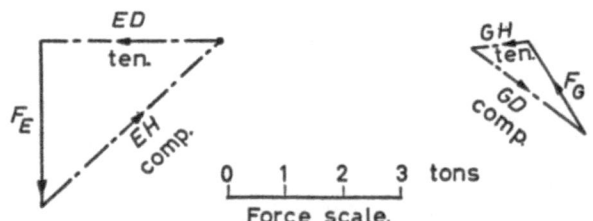

Figure 4.26. Equilibrium of joints E *and* F *from Figure 4.25*

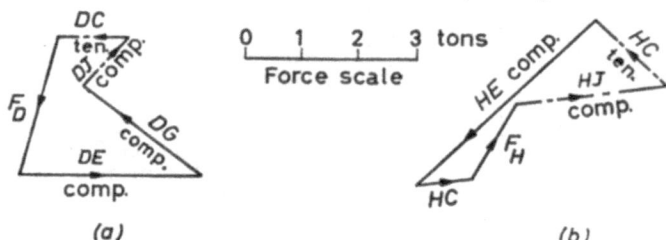

Figure 4.27. Equilibrium of joints D *and* H

116

The cantilever truss shown in *Figure 4.25* has six free joints. *It has 12 members and thus it is just stiff.*

Another way to judge whether a truss is just stiff would be to attempt to determine thrust forces in each member. If these can be

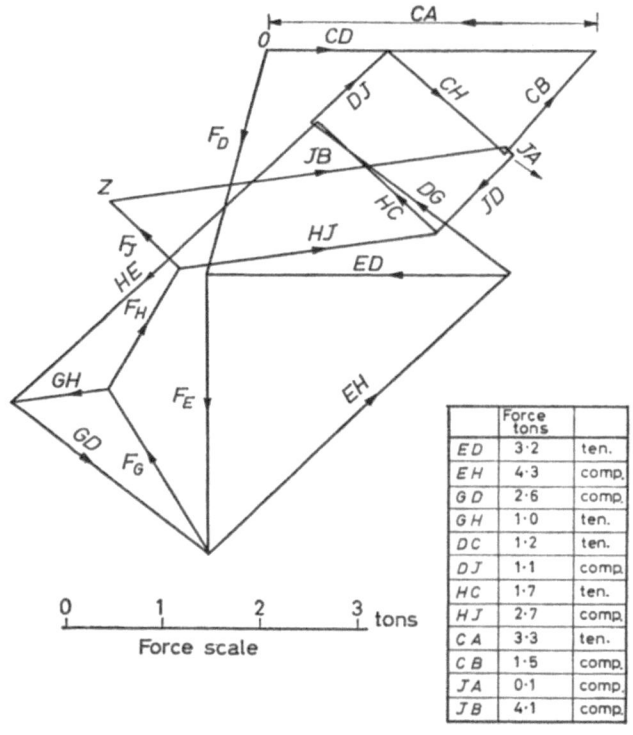

	Force tons	
E D	3·2	ten.
E H	4·3	comp.
G D	2·6	comp.
G H	1·0	ten.
D C	1·2	ten.
D J	1·1	comp.
H C	1·7	ten.
H J	2·7	comp.
C A	3·3	ten.
C B	1·5	comp.
J A	0·1	comp.
J B	4·1	comp.

Figure 4.28. Stress diagram of Example 4.10

found from conditions of equilibrium, the structure is stable and determinate.

In *Figure 4.25*, loads are acting on hinge points. If we take point *E*,

117

the acting load can be resolved into directions *ED* and *EG*. This can be carried out graphically as in *Figure 4.26(a)*.

The triangle of forces at hinge *E* shows *ED* in tension (arrow pointing away from hinge) and *EH* in compression (member pressing on hinge). Similarly, in *Figure 4.26(b)*, a resolution of forces at hinge *G* shows *GD* in compression and *GH* in tension.

It will then be possible to progress to hinges *D* and *H* and solve these in the same manner: at *D*, the unknown members *DC* and *DJ* will be found and at *H*, members *HC* and *HJ*. These graphical solutions are shown in *Figure 4.27(a)* and (*b*). Note that values of *ED*, *EH*, *GD* and *GH* were taken from *Figure 4.26*.

The process can be repeated for *C* and *J* and thus the forces in *CA*, *CB*, *JA* and *JB* will be found. This has not been done and is left as an exercise for the reader.

Note that the reaction at *A*, R_A, holds *AC* and *AJ* together in equilibrium and similarly for R_B which balances *BC* and *BJ*. R_A and R_B are thus determined by the geometry of the truss and may be found by drawing vector triangles for points *A* and *B*.

The entire operation can be carried out in a comprehensive *stress diagram* in *Figure 4.28*. In this, external forces are plotted in a sequence from *C* to *J*, around the structure, and vector triangles are drawn into this hinge by hinge, utilizing solutions of previous hinges as we go along.

This completes Example 4.10.

4.6. STABILITY OF SPACE TRUSSES

The argument of *Figure 4.19* still holds for a space truss: as long as the joints at the ends of articulated members *allow rotation in all directions* and (space) *forces are acting only on the joints*, the members *carry only axial thrust*.

Figure 4.29(a) shows the basic space truss. It has been formed from a plane truss, *ABC*, with the addition of one hinge, *D*, out of the plane.

The basic space truss has four hinges and *six* members. Any additional hinge requires *three* members to attach it to the basic form, as *E* in *Figure 4.29(b)*.

A just stiff space truss will have

$q = 3j - 6$ members (three members for each joint, less the six members of the basic form).

Since the structure would have j_s solid supports which do not need to be attached by additional members,

$$q = 3j - 6 - (3j_s - 6) = 3(j - j_s).$$

The number of members required to make a space truss just stiff is *three times the number of free joints*.

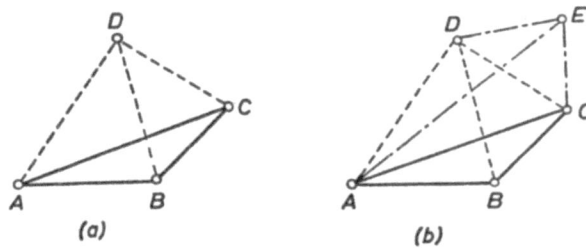

Figure 4.29. Building up of a space truss form

Example 4.11

Figure 4.30. Just stiff prism of Example 4.11

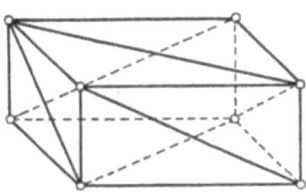

The rectangular (unsupported) prism of *Figure 4.30* is just stiff because it has

$$q = 3 \times 8 - 6 = 18 \text{ members.}$$

Example 4.12

The pyramid of *Figure 4.31* is once indeterminate, since

$$q = 3 \times \text{(free joints)} = 3 \times 1 = 3.$$

Any one of the bars connecting E to the base may be omitted and the structure will then be stable and determinate.

119

We may look at this problem in the same way as Example 4.10 was considered. External forces will act at E, and resolution along three

Figure 4.31. Statically indeterminate pyramid

directions in space would allow determination of *three* unknown member forces meeting at E *but not more than three*. Having four unknown members means one redundancy in this case.

Example 4.13

This will be an exercise in the simple use of the principle of equilibrium to calculate forces in members of a space truss. A joint is in equilibrium under the combined effect of the external forces acting on it and the axial thrust forces in the members meeting there. Each of these forces will have components along the x, y and z axes and $\Sigma F_x = 0$, $\Sigma F_y = 0$, $\Sigma F_z = 0$ will provide three equations from which three unknown member forces can be found.

To determine force *components*, place the co-ordinate system with its *origin at the joint*. Assume *all members in tension*, that is, pointing away from the joint. Then F_x for any force $= F.\cos \alpha_x$ where

$\cos \alpha_x$, the x-direction cosine for the force, is $\frac{x}{L}$. Here L is the *true length* of a given member and x is the ordinate of the length, usually read from a 3-projection drawing. F_y and F_z are formed similarly.

Since all forces are assumed to point away from the joint, the sign of the components F_x will be positive when this coincides with the $+x$ axis.

Figure 4.32 shows a mast (a pyramid) of three members—it is just stiff. This also means that forces in members will be found by applying conditions of equilibrium to joint D.

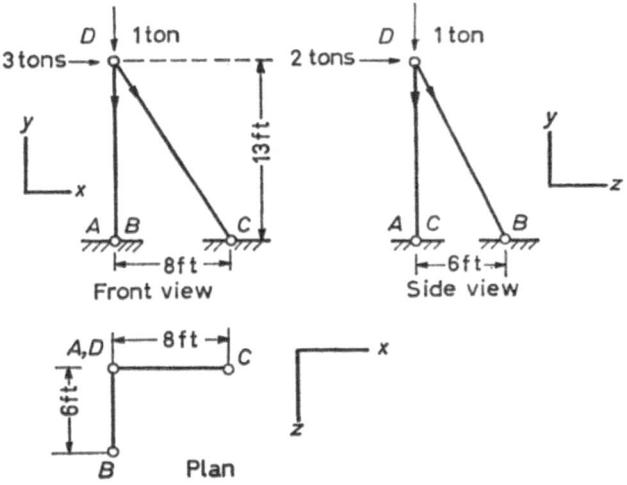

Figure 4.32. Mast of Example 4.13

Table 4.2. Equilibrium of joint D, Example 4.13

	F_x	F_y	F_z
1 ton	0	-1	0
2 tons	0	0	$+2$
3 tons	$+3$	0	0
F_{DA}	0	$-F_{DA}$	0
F_{DB}	0	$\dfrac{-13}{14\cdot3} F_{DB}$	$\dfrac{+6}{14\cdot3} F_{DB}$
F_{DC}	$\dfrac{+8}{15\cdot3} F_{DC}$	$\dfrac{-13}{15\cdot3} F_{DC}$	0

Totals: see below.

DC: actual length $= \sqrt{13^2 + 8^2} = 15\cdot3$ ft
DB: actual length $= \sqrt{13^2 + 6^2} = 14\cdot3$ ft
Assume DA, DB, DC all in tension.

121

Note: Member DA coincides with the $-y$-axis, so that $\frac{y}{L} = -1$.

x: $+3 + 0.52F_{DC} = 0$ or $F_{DC} = -5.7$ tons (comp.)

y: $-1 - F_{DA} - 0.91F_{DB} - 0.85F_{DC} = 0$

z: $+2 + 0.42F_{DB} = 0$ or $F_{DB} = -4.8$ tons (comp.)

and, substituting into y:

$$-1 - F_{DA} - 0.91\,(-4.8) - 0.85\,(-5.7) = 0$$
$$-1 - F_{DA} + 4.4 + 4.9 = 0$$
$$F_{DA} = 8.3 \text{ tons (tension).}$$

Incidentally, reactions at A, B and C have also been found since R_A is equal and opposite to F_{DA} and this applies to the other supports in the same manner.

This completes Example 4.13. This method will be further explained in Chapter 5.

4.7. ARTICULATED FRAMES

In Section 4.3, the three-hinged frame was discussed as a simple form of a statically indeterminate structure which was made determinate (and still remained stable) by the insertion of a hinge.

The same process may be used to form determinate shapes from other, more complex structures.

As a general rule, members which have external loads acting along their length, will normally have a bending moment and shearing force in their sections, as well as axial thrust. Members, hinged at their ends, which receive their load through the end hinges, will only have axial thrust. Thus the structure will be made up as a composite of beam-type elements and truss-type elements.

Many mechanical devices and machines are articulated frames. The simplest are a pair of scissors or pliers; more complex are cranes or aircraft landing gear. The toggle of Example 4.4 is, in reality, an articulated frame: member BD is the beam member and link AC is the truss-type member which carries only direct tension.

Example 4.14

In *Figure 4.33(a)* the three-hinged frame *ACB* is obviously unstable, with one degree of freedom. Roller *A* will move to the left and the structure will collapse. In order to make the frame stable, we pin the tie member *DE* on the frame. *AC* and *CB* are still beam members; *DE* is merely hinged onto them.

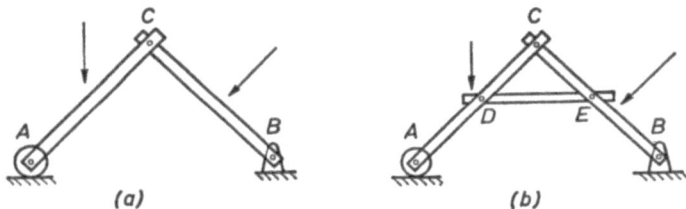

Figure 4.33. Three-hinged frames. (a) Unstable. (b) Tied, stable

In the case of the three-hinged frame of *Figure 4.10(b)*, the four reaction components (two at *A* and two at *B*) were found from the three conditions of planar equilibrium and the additional condition that bending moment at the apex hinge *C* was zero. In *Figure 4.33(b)* there are apparently only three unknown reaction components (one at *A*, two at *B*). However, the force in the tie *DE* is the fourth unknown; the three conditions of equilibrium and the zero bending moment at *C* still supply sufficient equations to solve for all unknowns.

Note that points *D* and *E* are not articulated, only the tie *DE* is hinged (hung) onto the beam members *AC* and *CB*.

Example 4.15

Frame *ACDB* has four hinges and will collapse, by *B* moving to the right. To steady the frame, knee ties *EG* and *HJ* were inserted and pinned to the main members.

It is easy to see that each hinge requires one tie, placed in a position where it will steady the two members connected by the hinge. This rule holds for any number of hinges.

Since the frame as a whole is in equilibrium, R_A and R_B are each 2 tons, vertical upwards. This will enable us to determine internal forces throughout *AC*, as shown in *Figure 4.34(b)*.

123

From the geometry of AC:

$$A'C = \frac{10}{2} = 5 \text{ ft}$$

and

$$CK = \frac{2}{2} = 1 \text{ ft}$$

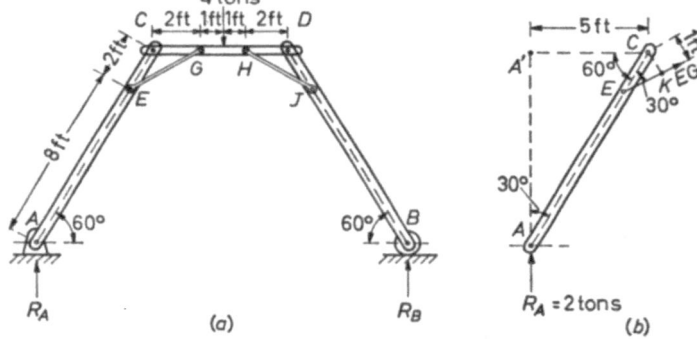

Figure 4.34. Four-hinged frame of Example 4.15

Then bending moment at C:

$$BM_0 = R_A \times A'C - EG \times CK = 2 \text{ tons} \times 5 \text{ ft} - EG \times 1 \text{ ft} = 0$$

that is, $EG = 10$ tons, tension as assumed.

All internal forces may now be found, by taking each of the frame members AC, CD and DB separately.

This completes Example 4.15.

4.8. SUMMARY: STABILITY AND STATICAL DETERMINACY

Stability results from adequate support action. Generally, a space body requires six, and a plane body three reaction components to be stable. More restraints make the structure redundant.

Redundant bodies may be made determinate by articulation. The insertion of hinges into redundant frames or beams can result in stability and determinacy.

124

Structures may be built up out of single straight members, by joining them together by means of hinges and applying loads to these joints only. These trusses will carry axial thrust only in the members and will generally be stable and determinate if the number of members is twice the unsupported joints in a plane truss and three times the unsupported joints in a space truss.

A test for stability and determinacy is whether internal forces can be found by applying conditions of equilibrium. Reactions may be found direct or by working from joint to joint through an articulated structure towards the support hinges. Whenever the structure is stable and determinate, reactions and forces in members have a single value for a given loading and this value can be found from the conditions of equilibrium.

4.9. Problems

In answers, + or − signs mean that components are in the + or − directions of co-ordinate axes. It is advisable to assume, for a start, reaction components acting along + co-ordinate directions. Then a *positive result* will show that the direction was correctly assumed; it is also pointing along the + axis!

Problem 4.1

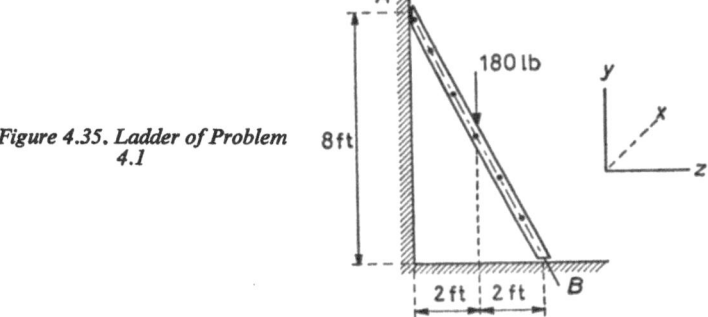

Figure 4.35. Ladder of Problem 4.1

A man, weighing 180 lb, is working in a position half-way up the ladder of *Figure 4.35*. Assuming that the reaction at *A* is at right angles to the wall, determine the minimum value of the coefficient of static sliding friction at *B* required for stability.

Note: The coefficient of friction, $f = \dfrac{F_{Bz}}{F_{By}}$

ANS: $f = \frac{1}{4}$

Problem 4.2

The structure of *Figure 4.36* is hinged to a solid support at *A* and *B*. The support at *B* is arranged in such a manner that the reaction F_B makes 45 degrees with the horizontal. What are the reaction components and the axial thrust forces in the structure?

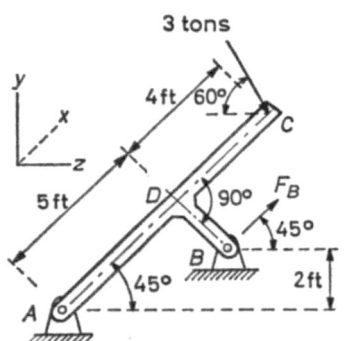

Figure 4.36. Structure of Problem 4.2

ANS: $F_{Ay} = -5 \cdot 9$ tons; $F_{Az} = -9 \cdot 9$ tons
$F_{By} = F_{Bz} = +8 \cdot 5$ tons
$AD = 11 \cdot 3$ tons (tension)
$DC = 0 \cdot 8$ ton (compression)
$DB = 0$

Problem 4.3

The space frame *ABCD* of *Figure 4.37* supports a flagpole at *A* and the applied loads there are as shown. Find the axial thrust forces in all members. Adopt a suitable system of co-ordinates.

ANS: *AB* 1,060 lb (tension)
AC 640 lb (compression)
AD 39 lb (tension)

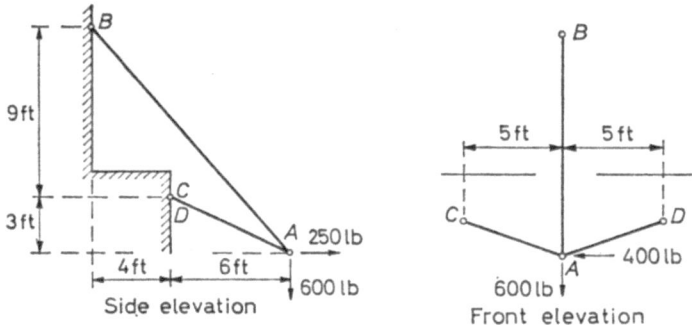

Figure 4.37. Space frame of Problem 4.3

Problem 4.4

A saw-tooth roof frame shown in *Figure 4.38* is hinged to solid supports at A and B. Find components of reactions at A and B when the two loads shown are acting.

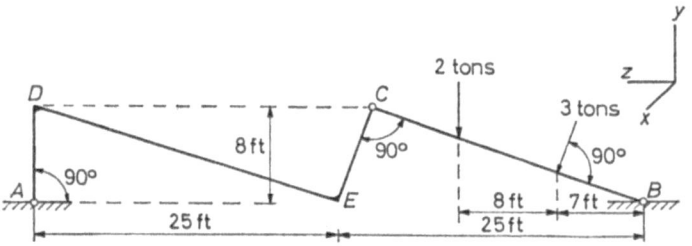

Figure 4.38. Saw-tooth frame of Problem 4.4

ANS: $F_{Ay} = +1 \cdot 04$ tons; $F_{Az} = -3 \cdot 66$ tons
$F_{By} = +3 \cdot 78$ tons; $F_{Bz} = +2 \cdot 64$ tons

(Note: The line of the reaction F_A will pass through hinge C. This happens because there is no acting load on $ADEC$ and the only force on this part of the structure is F_A. Since the bending moment at C is zero, and F_A is not zero, its distance from C must be zero. This provides an easy approach graphically. It is advisable, however, to check results by calculation.)

127

Problem 4.5

Find all reaction components at *A* and *B* for the **Gerber**-type beam shown in *Figure 4.39*.

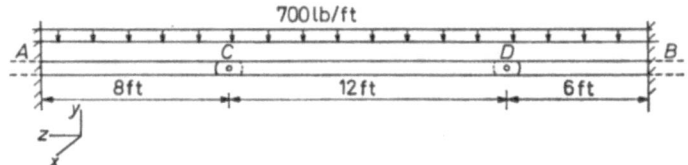

Figure 4.39. Gerber-beam of Problem 4.5

$$\text{ANS: } F_{Ay} = +9,800 \text{ lb}; \quad M_{Ax} = -56,000 \text{ lb-ft}$$
$$F_{By} = +8,400 \text{ lb}; \quad M_{Bx} = +37,800 \text{ lb-ft}$$

(Note: If hinges *C* and *D* were omitted, on the fully encased beam *AB* the reaction components would be:

$$F_{Ay} = F_{By} = +9,100 \text{ lb};$$
$$M_{Ax} = M_{Bx} = +39,500 \text{ lb-ft}$$

These results were obtained by using methods of elasticity.)

Problem 4.6

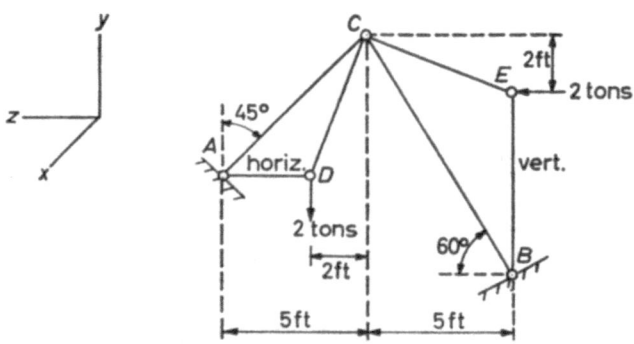

Figure 4.40. Frame of Problem 4.6

The plane structure of *Figure 4.40* is hinged to rigid bases at *A* and *B*. It may be regarded as a fully articulated truss with three free joints C, D and E, *or* as a three-hinged frame with hinges at A, B and C.

(*a*) Is the structure stable under any plane loading?
(*b*) What are the reaction components at A and B?

> ANS: (*a*) Yes. The structure is just stiff
> (*b*) $F_{Ay} = +2.21$ tons; $F_{Az} = -1.41$ tons
> $F_{By} = -0.21$ ton; $F_{Bz} = -0.59$ ton

Problem 4.7

In *Figure 4.41(a)* and (*b*) a plane truss of rectangular shape is supported at A and B. In both cases A is a plane truss hinge. In diagram (*a*), however, B is a plane roller and in (*b*) it is a plane truss hinge. In (*b*) the diagonal member CD is missing.

Are both trusses just stiff?

Determine reaction components at A and B for truss (*a*).

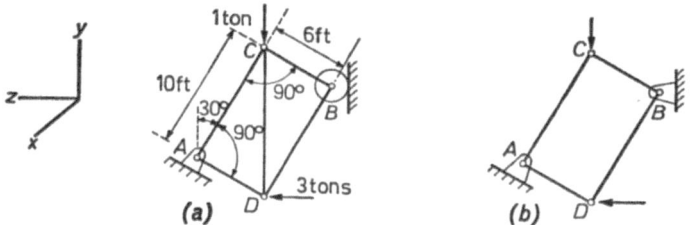

Figure 4.41. Rectangular trusses of Problem 4.7

> ANS: Yes. In (*a*), support B is not a *rigid* support and the extra member CD is required to compensate for the missing component (restraint) at B.
>
> $F_{Ay} = +1$ ton; $F_{Az} = -5.47$ tons; $F_{Bz} = +2.47$ tons

Problem 4.8

The tower structure shown in *Figure 4.42* has hinged joints. Is it just stiff? How many additional members would it need for a third storey, 12 ft-square in plan, 10 ft above frame $KLMN$?

> ANS: Yes. Eight free joints and twenty-four members make it just stiff. Four extra joints will require another twelve members.

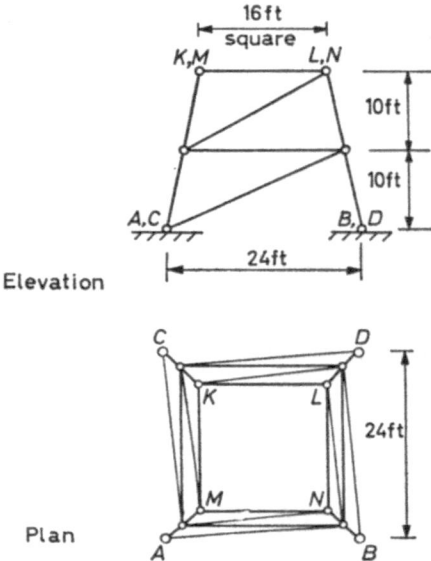

Figure 4.42. Tower of Problem 4.8

Problem 4.9

A cube of 6 in edge length is suspended by means of three links as illustrated in *Figure 4.43*. What are the forces in the links?

(Hint: All forces meet at point D. From the conditions of equilibrium for D in the form of $F_{Dx} = F_{Dy} = F_{Dz} = 0$, the three link forces can be found.)

> ANS: $AE = 147$ lb tension
> $BF = 22$ lb tension
> $CG = 195$ lb tension

Problem 4.10

The three-hinged roof frame ACB has a 'collar-tie' DE to prevent the roller support A from moving out sideways, as shown in *Figure 4.44*. Find the force in DE.

> ANS: 4,200 lb tension

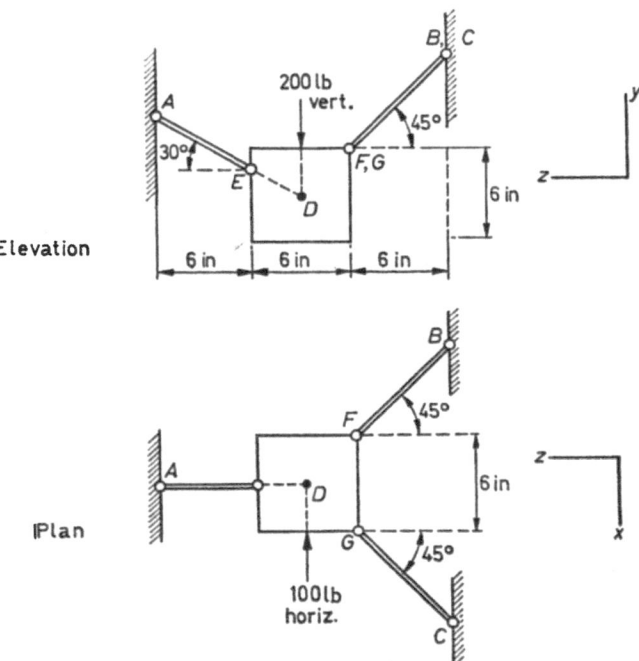

Elevation

Plan

Figure 4.43. Link supported cube of Problem 4.9

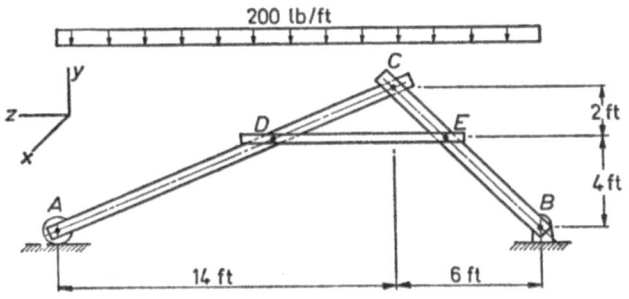

Figure 4.44. Roof frame of Problem 4.10

131

Problem 4.11

Figure 4.45 shows a Gerber-beam with fully encased ends. If we remove hinges C and D, the restraining moments will be $M_{Ax} = -\dfrac{w.L^2}{12}$ and $M_{Bx} = +\dfrac{w.L^2}{12}$, derived from using principles of elasticity. How much should be the distance $AC = DB$ in order to have the same end moments on the Gerber-beam?

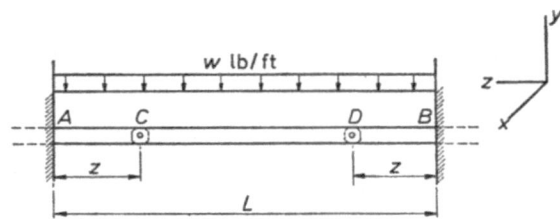

Figure 4.45. Gerber-beam of Problem 4.45

ANS: $z = 0\cdot21\ L$

Problem 4.12

Determine the distance $AC = DB$ in Problem 4.11 if instead of a uniformly distributed load, the beam is loaded with a solitary midpoint concentrated load F. The restraining end moments for this case, with hinges C and D removed, are

$$M_{Ax} = -\frac{FL}{8} \quad \text{and} \quad M_{Bx} = +\frac{FL}{8}$$

ANS: $z = 0\cdot25\ L$

132

CHAPTER 5

EQUILIBRIUM OF PARTS

In Chapter 2, we have seen how to combine and resolve forces and how to transpose force systems. In Chapters 3 and 4 the conditions of stability and equilibrium were discussed. Support (reaction) forces were calculated in systems which were stable and statically determinate.

In these 'just stiff' systems, then, *all* external forces (acting and reaction forces) will be known or calculated. The system *as a whole* is in *equilibrium* under the action of this total external force system.

For a structure to be designed, it is essential that *internal forces throughout the structure* should also be known. This chapter will deal with internal forces and the means whereby the conditions of equilibrium can be applied to the problem of finding internal forces.

It should be emphasized, however, that *reactions must be determined first*, and *before* any attempts are made to find internal forces. There are exceptions to this rule (in the case of cantilevers, for instance) but generally, nothing can be done *until all external forces, including reactions*, are known.

This condition applies to *any* stable structure. When the body has too many restraints (or too many members) and therefore, it is redundant, the reactions will be determined by applying the rules of Elasticity. Once reactions are known, the process of finding internal forces will generally be the same as for a statically determinate body.

It will help to re-state this all-important sequence of 'solving' any stable structure:

Step 1. Establish geometry of body: dimensions of structure and types of restraint at supports.

Step 2. Locate all *acting* forces: position; line of action and direction; magnitude.

Step 3. Determine *all* reaction components at *every* support.

Step 4. Find *all* internal force components in every section of the body.

The concept 'force' naturally includes both loads and moments.

Step 3 will be accomplished by using the method of transposition, transposing to the points of support in a systematic fashion. If the structure is determinate, sufficient equations will be available to calculate all the unknown reaction components throughout the structure. If the body is redundant, there will be fewer equations than unknowns and an elastic analysis must be used which we do not discuss in this book.

Step 4 can only start after Steps 1, 2 and 3 have all been completed. (The exception is the case of a cantilever; here we do not require the reactions.) The conditions of equilibrium will now be applied to a part of the body and the internal forces will be found at the section bounding the part.

Before we embark on this, we shall state it once more: we *must establish the equilibrium of the body as a whole* (under acting and reaction forces), before the equilibrium of any of its portions can be analysed.

5.2. INTERNAL FORCES: DEFINITION AND CONVENTIONS

The Co-ordinate System

The concept of *internal forces* was explained and illustrated in Section 1.4. Further reference was made to them when the method of transposition was introduced in Chapter 2. Looking back to Example 2.4 and *Figure 2.21*, we are reminded of the *six components* which make up the internal forces in a body. Let us have another look at this, in *Figure 5.1(a)*.

AB in *Figure 5.1(a)* is a rigid body in equilibrium. The bodies that will be considered are long compared to their thickness and it is usual to take, at least for calculations in Statics, the 'centroid-line' as representing the 'line' of the structure. The centroid-line connects the centroids of successive cross-sections along the body. A cross-section (or simply, *section*) then is defined as a cut through the body *at right angles to the centroid-line*. *Figure 5.1(b)* shows the cross-section at *C* with a short portion of the centroid-line; this portion, according to our definition, is at right angles to the paper, at least in the close vicinity of the section.

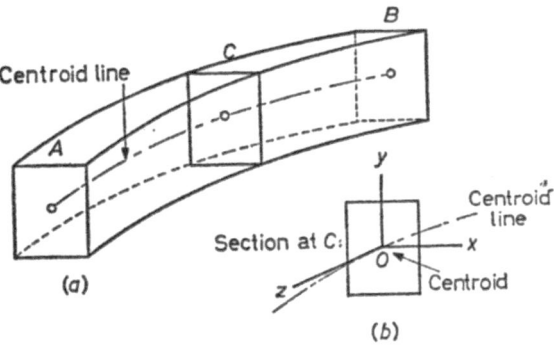

Figure 5.1. The standard system of axes

Internal forces at the section will be defined with reference to a co-ordinate system, the origin of which is at the centroid *O*. The axis *z* will be *along* the centroid-line and the *y* and *x* axes will coincide with the *principal axes* of the section. Whenever the section has an axis of symmetry, one co-ordinate axis, say, *y*, is made to coincide with it. The other, in this case *x*, will then start at *O* and be drawn at right angles to *y*. This pair of axes will then be the principal axes. If the section has no axis of symmetry, the position of principal axes can be calculated: the two principal positions for *x* and *y* will be those axes for which the second moment of area (moment of inertia) of the section

135

has maxima and minima values, respectively. It is assumed that the student is familiar with the concepts of second moment of area and principal axes. The textbook: *Engineering Mechanics* by Timoshenko and Young (McGraw-Hill), Part 1: Statics, deals with these concepts in its Appendix 1 in considerable detail.

A solid or hollow circular section is a special case since *any* pair of mutually perpendicular axes may be adopted for x and y, all are principal axes. In such a case one axis, say y, is adopted in the direction (or in the plane) of the majority of the.

Table 5.1. Typical sections and usual choice of x and y axes

Symmetrical sections (at least one axis of symmetry)		Non-symmetrical sections
		x and y are not principal axes:

loads on the structure and x will then pass through centre O of the circle, at right angles to y. For structures loaded vertically, y is vertical and x is horizontal.

An arbitrary choice for y and x is sometimes adopted for non-symmetrical extruded sections such as angles. It is customary to take y as parallel to one leg of the angle and x as parallel to the other, both through the centroid O. Tables of sections list properties of these angle sections on this basis and it is understood that such pairs of axes are *not* principal axes of the section. Table 5.1 illustrates some sections and the choice of x and y.

If we wish to adhere to the rules and signs of transposition as discussed in Chapter 2 (refer to *Figure 2.19*), we adopt a *right-hand screw* co-ordinate system. That is, the positive directions of the co-ordinates will be as shown in *Figure 5.1(b)*. In a right-hand screw system, $+x$ can be turned into $+y$ by anti-clockwise rotation, when viewed from the $+z$ direction.

In this system, the *positive z-component* of a force is directed *from O along the positive z-axis*. Positive x and y components are similarly defined. A *positive rotation* about the z-axis means that the rotation is *clockwise* when viewed *from* the positive end of the z-axis. This rule also applies to rotations (moments) around the x and y axes.

Internal forces

In *Figure 5.2*, we reproduced portion AC of the rigid body AB from *Figure 5.1(a)*. In *Figure 5.2* it is assumed that the body is supported at A and that various loads and moments are acting on the portion AC.

AB as a whole is in equilibrium, under all acting and reaction forces. AC is also in equilibrium and this equilibrium is established by considering on the one hand, the external forces acting on the portion AC, and on the other, *the internal resistance (strength) of section C*. For AC to be in equilibrium (to remain, in fact, part of the solid body AB), it is required that the *resultant of the external forces on AC* and the *internal resistance*

137

on section C shall be in equilibrium together. In the terminology of Chapter 2, the resistance at *C* is the *balancing force* of the *resultant of loads and reactions on portion AC.*

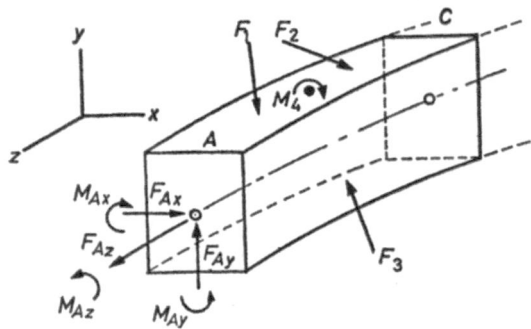

Figure 5.2. Concept of internal forces

In vector terms:

$M_A(x, y, z) + F_A(x, y, z) + F_1 + F_2 + F_3 + M_4 =$ Resistance of section *C*

This equilibrium can be established *at* section *C* by transposing the entire load system of *AC* (including reactions) to *C*. The resultant thus obtained will then represent the local effect of loads at *C* and the material of the section must be able to exert an exactly equal and opposite resistance, balancing the loads.

The 'internal force' at a section may be defined as the *resultant* of loads transposed to *C or* as the numerically equal, but in sign opposite, *balancing* force. In either case, the material of the section, its geometry etc., must be designed in such a way that the section will carry the internal force safely.

For the purpose of this chapter, the internal force will be defined as *the resultant* of loads (including reactions) on the portion to one side of the section considered. This means, that as far as signs are concerned, these will be according to the signs of the components $R(x, y, z)$ and $M(x, y, z)$ of the resultant of the force system on AC, transposed to C.

In Example 2.4, loads were transposed to section O of a cantilever beam. *Figure 2.21* which shows this structure, is reproduced in *Figure 5.3(a)*.

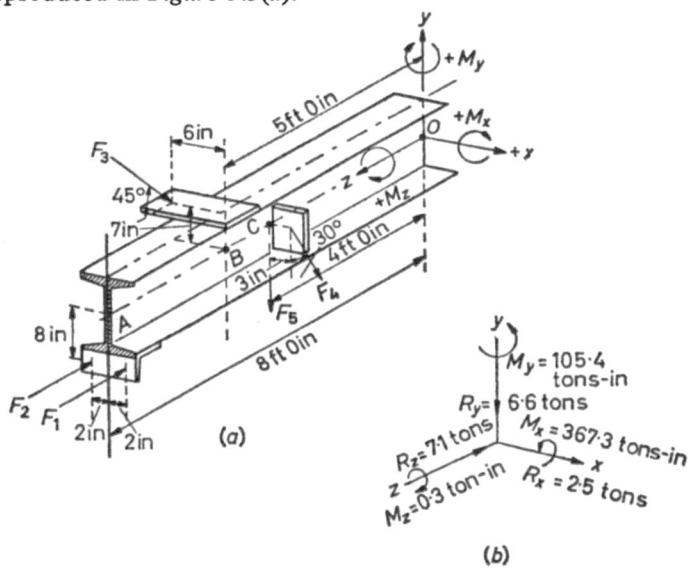

Figure 5.3. Cantilever steel beam of Example 2.4

Components of the transposed resultant at O (as obtained from Table 2.2) are listed in Table 5.2. The components were plotted in *Figure 5.3(b)*.

It is possible to combine these components into a pair of R and M vectors but seldom is anything gained by this. Design codes usually specify required material properties and permissible stresses according to one or more of the *components* of the internal force.

139

Table 5.2. Components of internal force at O in *Figure 5.3(a)*

R tons	x	$+2 \cdot 5$	Horizontal (lateral) shearing force
	y	$-6 \cdot 6$	Vertical shearing force
	z	$-7 \cdot 1$	Axial thrust force (here, compression)
M tons-in	x	$-367 \cdot 3$	Longitudinal (beam) bending moment
	y	$-105 \cdot 4$	Horizontal (lateral) bending moment
	z	$+0 \cdot 3$	Twisting (torsional) moment or torque

Internal force diagrams

The structure will be finally solved if internal forces are fully known in *every* section. This may mean proceeding from section to section, at intervals of, perhaps, a few inches at a time. At every section the components of R and M would be calculated by transposition.

In most cases the work involved can be reduced by calculating the internal forces only at a few key points. In between these points, variation of a given internal force may then be described by an equation or by plotting the equation in the form of a diagram.

Six diagrams will show the variation of each of the six internal force components and will tell the designer at a glance where the internal forces reach maxima or critical values for the design and conversely, where the internal force values are low—here splices or joints can be most conveniently located.

5.3. CONCENTRATED LOADS

The principle and rules of plotting internal force diagrams can be shown by the example of a simply supported *plane* beam with a single mid-point *concentrated* load. In this case, shown in *Figure 5.4(a)*, there will only be three internal force components anywhere on the beam:

$$R_y \; - \text{(vertical) shearing force}$$
$$R_z \; - \text{axial thrust and}$$
$$M_x \; - \text{(longitudinal) bending moment.}$$

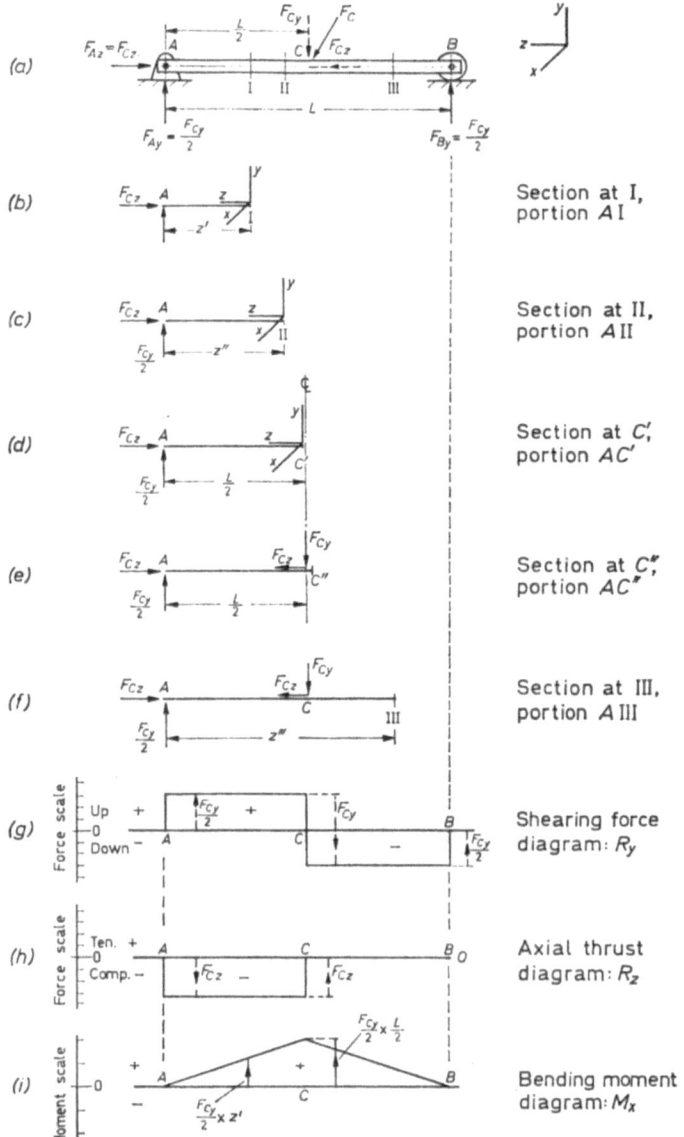

Figure 5.4. Derivation of internal force diagrams for concentrated loading

In choosing key sections, some experimenting may be necessary. After finding the reactions, we have established equilibrium and can look for a suitable section to start with. If we choose the arbitrary section I, at a distance z' from A, we find the following internal force components by inspection, *see Figure 5.4(b)*:

$$R_y = + \frac{F_{Cy}}{2} \text{ (i.e. up)} \qquad \text{Section at I}$$
$$R_z = - F_{Cz} \text{ (i.e. compression)} \qquad \text{Portion } AI$$
$$M_x = + \frac{F_{Cy}}{2} . z'$$

Next, we move on to section II which is at a distance z'' from A. The resultant at II of forces on AII has these components, as shown in *Figure 5.4(c)*:

$$R_y = + \frac{F_{Cy}}{2} \text{ (up)} \qquad \text{Section at II}$$
$$R_z = - F_{Cz} \text{ (compression)} \qquad \text{Portion } A\text{II}$$
$$M_x = + \frac{F_{Cy}}{2} . z''$$

If we compare the internal forces at I and II, we find that values of the shear force and axial thrust remained constant. This will still be true for z values up to $\frac{L}{2}$, in fact right up to the section C itself. So that if we take a section at C as in *Figure 5.4(d)*, in the immediate vicinity *left* of C, values of R_y and R_z are still the same.

Rule 1. On straight parts of the structure which are not loaded, the force components (shear and thrust) of the internal force are *the same* in every section.

Rule 2. On parts of the structure which are not loaded, the *moment* component (bending moment) varies *linearly* with the distance from the force.

142

The equation for this can be written, for portion AC:

$$M_x = + \frac{F_{Cy}}{2} \cdot z$$

On the basis of the foregoing, we can plot a diagram for each of the three components R_y, R_z and M_x for the part AC. This was carried out on the left-hand side of *Figures 5.4(g)*, *(h)* and *(i)*. A suitable force-scale was used for R_y and R_z (not necessarily the same for the two diagrams) and a moment scale for M_x.

Section III is on the right half of the beam, between C and B, see *Figure 5.4(f)*. By inspection the following components are obtained for Section III:

$$R_y = + \frac{F_{Cy}}{2} - F_{Cy} = - \frac{F_{Cy}}{2} \text{ (down)} \qquad \text{Section at III}$$

$$R_z = + F_{Cz} - F_{Cz} = 0 \qquad\qquad\qquad \text{Portion } A\text{III}$$

$$M_x = + \frac{F_{Cy}}{2} \cdot z''' - F_{Cy}\left(z''' - \frac{L}{2}\right) = + \frac{F_{Cy}}{2}(L - z''')$$

or at any section between C and B:

$$M_x = + \frac{F_{Cy}}{2}(L - z).$$

We note that the values for R_y and R_z are the same anywhere *on part CB*, from C'' to B, conforming to Rule 1. These constant values were plotted in *Figures 5.4(g)* and *(h)*. The value of M_x will still vary linearly, according to Rule 2, but it will now diminish since the effect of the load at C, F_{Cy}, is to reduce the positive rotation of the reaction at A. The straight declining line for CB was plotted in *Figure 5.4(i)*.

Finally, at B, all internal force components become zero. This is normally so at the *ends* of beams, since past an end section there cannot be resistance to loads.

Throughout our discussion of the beam of *Figure 5.4(a)* we

were advancing from A towards B. Since the equilibrium of parts must be valid no matter what the direction of our approach is, the *same results* should be obtainable when passing from *B to A*. We should, however, remember the equilibrium of the body as a whole: for equilibrium, the resultant on the right portion of the body must be equal and *opposite* to the resultant on the left portion. This can easily be verified by taking portions BIII and BII, shown on *Figure 5.4A*. Note opposite signs in all expressions.

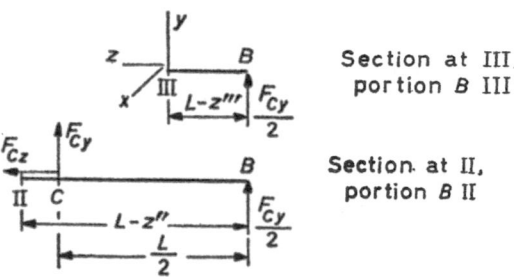

Section at III, portion B III

Section at II, portion B II

Figure 5.4A. The effect of reversal of portions

$$R_y = +\frac{F_{Cy}}{2} \text{ (up)} \qquad \text{Section at III}$$
$$R_z = 0 \qquad \text{Portion } B\text{III}$$
$$M_x = -\frac{F_{Cy}}{2}(L - z''')$$

$$R_y = +\frac{F_{Cy}}{2} - F_{Cy} = -\frac{F_{Cy}}{2} \text{ (down)} \qquad \text{Section at II}$$
$$R_z = +F_{Cz} \text{ (compression)} \qquad \text{Portion } B\text{II}$$
$$M_x = -\frac{F_{Cy}}{2}.(L - z'') + F_{Cy}\left(\frac{L}{2} - z''\right) = -\frac{F_{Cy}}{2}.z''$$

For several reasons, it is more convenient to proceed from left to right (or opposite the $+z$ direction). One is that the positive bending moment is traditionally the moment on a simply supported horizontal beam, under downward loads. This results in compression in the top half and tension in the bottom half of the beam. If we proceed from left to right, our sign convention will label this type of bending moment as positive, as shown in *Figure 5.4(i)*.

All this has shown one thing clearly: that *key sections* are at A, C (C' and C'') and B. If the internal force components are known in these key sections, the diagrams can be drawn out simply by connecting these values with straight lines.

In the case of a space structure loaded with concentrated loads, a similar reasoning will lead to six internal force diagrams, all made up of straight lines.

Example 5.1

For the cantilever steel beam of *Figure 5.3*, internal force diagrams are to be determined.

Key sections are A, B, C and O.

We reproduced here, from Chapter 2, Table 2.2, re-named here Table 5.3.

As we proceed from A towards O, relevant *portions* of this table may be used for calculating or checking internal forces. For instance, at Section A, F_1 and F_2 are only acting and

$$M_x = F_y.z - F_z.y = 0 - (+16 + 24) = -40 \text{ tons-in},$$

the same value as obtained from inspection (see below).

Section A. All components are zero (end of structure). However, immediately to the right of A, loads F_1 and F_2 will create internal forces which will be constant on AB:

$$R_z = F_{1z} + F_{2z} = -2 - 3 = -5 \text{ tons}$$
$$M_x = -(2 + 3) \times 8 \text{ in} = -40 \text{ tons-in}$$
$$M_y = -2 \times 2 \text{ in} + 3 \times 2 \text{ in} = +2 \text{ tons-in}$$

Section B. To the forces on AB is now added the effect of F_{3y} and F_{3z}:

$$\left. \begin{array}{l} R_y = F_{3y} = -2 \cdot 1 \text{ tons} \\ R_z = F_{1z} + F_{2z} + F_{3z} = -7 \cdot 1 \text{ tons} \end{array} \right\} \text{ on } BC$$
$$M_x, M_y\text{—as for Section } A.$$
$$M_z = F_x.y - F_y.x = 0 - F_{3y} \times 6 \text{ in} = +12 \cdot 6 \text{ tons-in}$$

Table 5.3. Example 5.1. Transposition to Section O

Forces	F_x	F_y	F_z	x	y	z	$F_{x.y}$	$F_{x.z}$	$F_{y.x}$	$F_{y.z}$	$F_{z.x}$	$F_{z.y}$
	tons			in			tons-in					
F_1	0	0	−2	+2	−8	+96	0	0	0	0	−4·0	+16·0
F_2	0	0	−3	−2	−8	+96	0	0	0	0	+6·0	+24·0
F_3	0	−2·1	−2·1	−6	+7	+60	0	0	+12·6	−126	+12·6	−14·7
F_4	+2·5	−4·3	0	+3	0	+48	0	+120	−12·9	−206	0	0
F_5	0	−0·2	0	0	0	+48	0	0	0	−10	0	0
Totals	+2·5	−6·6	−7·1				0	+120	−0·3	−342	+14·6	+25·3

Table 5.4. Example 5.1. Transposition to Section C

Forces	F_x	F_y	F_z	x	y	z	$F_{x.y}$	$F_{x.z}$	$F_{y.x}$	$F_{y.z}$	$F_{z.x}$	$F_{z.y}$
	tons			in			tons-in					
F_1	0	0	−2	+2	−8	+48	0	0	0	0	−4	+16
F_2	0	0	−3	−2	−8	+48	0	0	0	0	+6	+24
F_3	0	−2·1	−2·1	−6	+7	+12	0	0	+12·6	−25·2	+12·6	−14·7
F_4	+2·5	−4·3	0	+3	0	0	0	0	−12·9	0	0	0
F_5	0	−0·2	0	0	0	0	0	0	0	0	0	0
Totals	+2·5	−6·6	−2·1				0	0	−0·3	−25·2	+14·6	+25·3

Section C. Internal forces at *A* and *B* have been found by inspection. For *C* we will re-write the transposition table; for the five loads involved this is usually justified. Transposition will be made to *Section C*, with point *C* as the origin of the system of axes in Table 5.4.

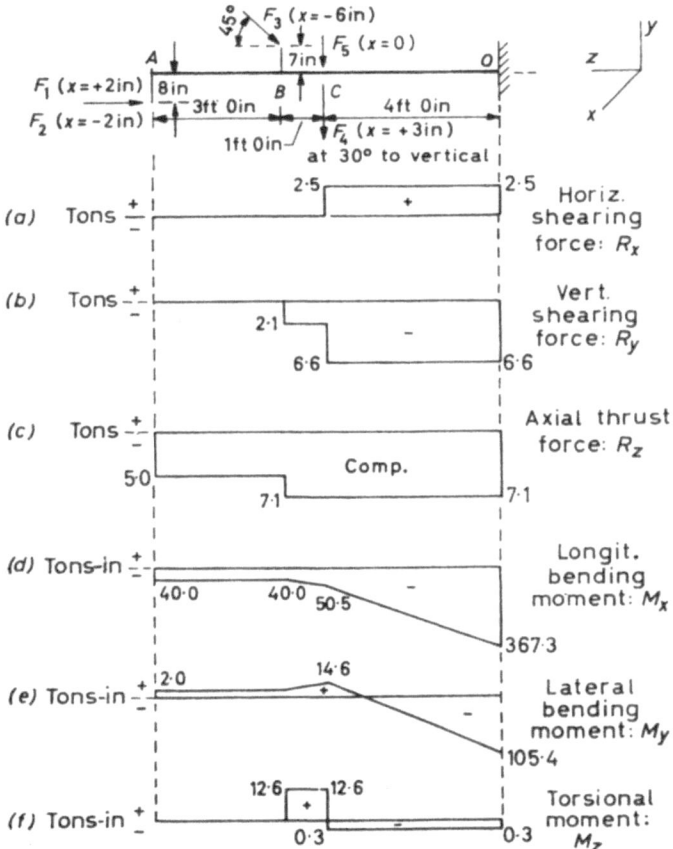

Figure 5.5. Internal force diagrams of Example 5.1

147

For Section C, then:

$$R_x = +2.5 \text{ tons}$$
$$R_y = -6.4 \text{ tons}$$
$$R_z = -7.1 \text{ tons}$$

$$M_x = -25.2 - 25.3 = -50.5 \text{ tons-in}$$
$$M_y = +14.6 - 0 \quad = +14.6 \text{ tons-in}$$
$$M_z = 0 - (-0.3) \quad = + 0.3 \text{ tons-in}$$

This completes calculations for the key sections A, B, C and O. From the values obtained, the six internal force diagrams were plotted in *Figure 5.5(a)* to *(f)*.

Notes on *Figure 5.5*:

We have followed Rule 1 in plotting R_x, R_y and R_z; also, Rule 2 in plotting the M diagrams. M_x is negative throughout which means that sections of the beam rotate in a manner which creates tension in its top half and compression in the bottom. M_y changes from positive to negative; when positive, the side on which F_4 is acting is in tension and this is changed to compression when M_y becomes negative.

The six diagrams of *Figure 5.5* supply complete information on internal forces throughout the beam. The designer, if it suits his work, may combine values of R_z, M_x and M_y to obtain 'fibre stresses'. Similarly, R_x, R_y and M_z can be combined in the calculation of 'shear stresses'.

The diagrams can be replaced by equations. It is also possible to feed the information of the transposition tables into a computer and obtain a 'print-out' of any variation of any of the six internal forces; in this way, the information can be stored on a punch card system.

5.4. UNIFORMLY DISTRIBUTED LOADS

In *Figure 5.6(a)* a plane beam is loaded with a 'uniformly distributed' load. This type of load can be envisaged as made up of a large number of individual concentrated loads of equal magnitude, ranged along the beam. For the calculation of reactions we may use the resultant of these since there we deal with the overall equilibrium of the beam. When analysing internal forces in sections, however, this is no longer possible and the amount of distributed load which is on *each portion* of the beam must be taken separately.

This being a plane problem, we only need to calculate R_y, R_z and M_x. In this case, however, $R_z = 0$ everywhere. Portion AI, *Figure 5.6(b)* has the following internal force components (by inspection):

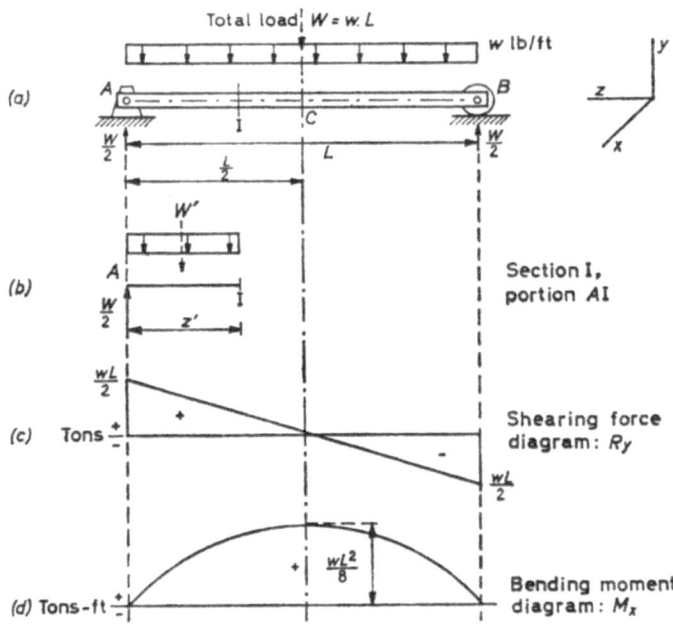

Figure 5.6. Derivation of internal force diagrams for distributed loading

$$R_y = +\frac{W}{2} - W' = +\frac{W}{2} - w.z' = +\frac{wL}{2} - wz' = w\left(\frac{L}{2} - z'\right)$$

$$M_x = +\frac{W}{2}.z' - W'.\frac{z'}{2} = \frac{wL}{2}z' - wz'.\frac{z'}{2} = \frac{w}{2}(Lz' - z'^2)$$

149

The equation of R_y is linear, that of M_x is quadratic. Generally, therefore, for any distance z from A:

$$R_y = w\left(\frac{L}{2} - z\right)$$

and

$$M_x = \frac{w}{2}(Lz - z^2)$$

These equations are plotted in *Figures 5.6(c)* and *(d)*. R_y is the (vertical) shearing force in a section and it becomes zero at the mid-point C, since for $z = \frac{L}{2}$:

$$R_y = w\left(\frac{L}{2} - \frac{L}{2}\right) = 0$$

At C, too, the parabola of the bending moment M_x has a maximum. The value of this maximum for $z = \frac{L}{2}$ is

$$M_x = \frac{w}{2}\left(L \cdot \frac{L}{2} - \frac{L^2}{4}\right) = \frac{wL^2}{8} = \frac{WL}{8}$$

We note here that a *zero value of the shearing force* coincides with the *maximum of the bending moment*. This finding has general validity and assists in finding the section where the bending moment has a critical value for design.

Rule 3. On straight parts of the structure which are loaded with uniformly distributed loading, the *force components* vary *linearly*.

Rule 4. On these parts, *moment components* vary as a *parabolic* curve.

Rule 5. For any type of loading, the maximum bending moment occurs when the corresponding shear force component is zero.

Example 5.2

In *Figure 5.7(a)*, a cantilever-type beam is loaded with a uniformly distributed load from end to end. Shearing force and bending moment diagrams are to be plotted.

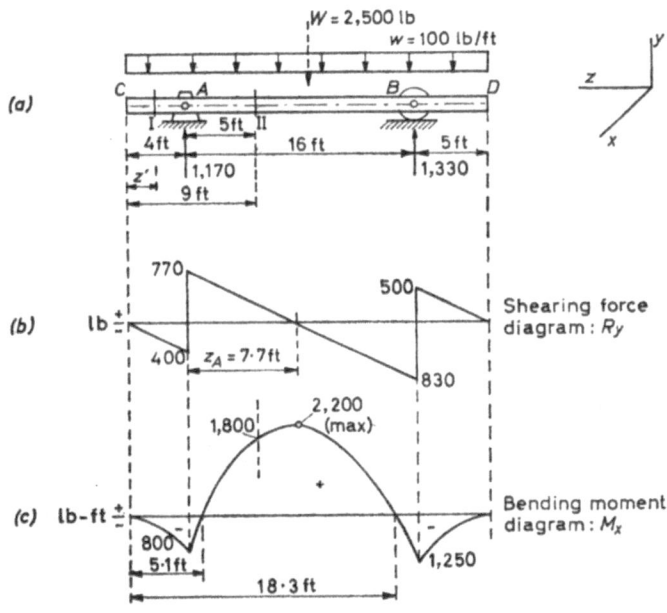

Figure 5.7. Example 5.2

To find F_A, we transpose to B. This operation can be simplified by summing all M_x values and equating to zero (since the beam as a whole is in equilibrium).

$$M_x \text{ at } B: \ -W \times 7\cdot5 \text{ ft} + F_A \times 16 \text{ ft} = 0$$

or $\qquad F_A = \dfrac{2{,}500 \times 7\cdot5}{16} = 1{,}170 \text{ lb (up, as assumed)}.$

For F_B, we may transpose to A in a similar simple fashion:

$$M_x \text{ at } A: + W \times 8.5 \text{ ft} - F_B \times 16 \text{ ft} = 0$$

that is, $\quad F_B = \dfrac{2,500 \times 8.5}{16} = 1,330 \text{ lb (up, as assumed)}.$

Since the body as a whole is in equilibrium, $R_y = 0$ and this provides a check for the calculated values of F_A and F_B:

$$R_y = F_A + F_B - W = 1,170 + 1,330 - 2,500 = 0. \text{ Correct.}$$

Plotting the two diagrams we start at the end C and note that both shearing force and bending moment are zero there. Moving towards A, we are accumulating a negative (down) shearing force and a negative M_x bending moment:

$$R_y = -100 \times z' \qquad \qquad \text{Section I}$$
$$M_x = -100 \times z' \times \frac{z'}{2} = -100 \times \frac{z'^2}{2} \quad \text{Portion } CI$$

Finally, at A (just to the left of A), the values are:

$$R_y = -100 \times 4 = -400 \text{ lb}$$
$$M_x = -100 \times \frac{4^2}{2} = -800 \text{ lb-ft}$$

Passing through A and in its immediate vicinity to the right:

$$R_y = -400 + F_A = -400 + 1,170 = +770 \text{ lb}$$

This operation was performed graphically in *Figure 5.7(b)* by plotting up (in the positive y direction) the 1,170 lb force occurring as a point load (reaction) at A. Incidentally, the shearing force diagram can be plotted in this way without much calculation. Starting at the left end of the structure, we plot loads as they are acting: a distributed load—as a straight line *sloping at the rate of loading* and a concentrated load *as a step* at the point where it occurs. At the end of the plot, in this case at D, we return to zero. This is also a check on our construction of the diagram.

For further plotting of M_x, we take Section II (portion CII). Section II is 9 ft from C.

$$M_x = -W'' \times 4.5 \text{ ft} + F_A \times 5 \text{ ft}$$
$$= -900 \times 4.5 + 1,170 \times 5 = +1,800 \text{ lb-ft}$$

This shows that the bending moment has changed from its negative value on CA to a positive value at II. Further plotting of points reveals that a considerable portion of the 'suspended span' AB has positive bending moment values. The change-over points where negative rotation gives place to positive rotation, will be found by writing out the equation of M_x in terms of the unknown distance, z: equating M_x to zero and solving for z will give this point, known as the point of contraflexure.

$$M_x = -(z.100).\frac{z}{2} + 1,170(z - 4) = 0$$
$$-50z^2 + 1,170z - 4,680 = 0; \text{ or, dividing by } -50:$$
$$z^2 - 23 \cdot 4z + 93 \cdot 6 = 0$$
$$z = \frac{+23 \cdot 4 \pm \sqrt{23 \cdot 4^2 - 374 \cdot 4}}{2} = \frac{+23 \cdot 4 \pm 13 \cdot 2}{2} = 18 \cdot 3 \text{ ft or } 5 \cdot 1 \text{ ft}$$

Finally, the bending moment at B may be found by considering the portion DB, approaching B from the right. Remembering that in this case signs will reverse,

$$M_x = -500 \times 2 \cdot 5 \text{ ft} = -1,250 \text{ lb-ft}$$

The maximum positive bending moment occurs where the shearing force is zero on AB. This occurs at a distance z_A from A.

$$z_A = \frac{770}{w} = \frac{770}{100} = 7 \cdot 7 \text{ ft}$$

The positive M_x maximum then is

$$-11 \cdot 7 \times 100 \times \frac{11 \cdot 7}{2} + 1,170 \times 7 \cdot 7 = +2,200 \text{ lb-ft}$$

The greatest *negative* M_x value occurs at B and there its value is *−1,250 lb-ft*. Note that the shearing force diagram passes through the zero line at three points: at A, at z_A from A, and at B. These are all 'maxima' values.

This completes Example 5.2.

5.5. MISCELLANEOUS AND MIXED LOADINGS

In addition to simple, concentrated and uniformly distributed loads, there is a variety of possible other loadings.

The *triangular* distributed load has varying rate or intensity of loading. The rate of loading varies *linearly* with distance along the beam. It can be shown that the shearing force line for this type of load is a parabola and the bending moment line is a cubic. Such loading occurs on the walls of water tanks or earth retaining structures.

The *moment-load* on a plane beam is a bending *load M_x*, that is, an applied rotation about the axis x which is at right angles to the plane of the beam. It causes a constant shearing force line and a linear bending moment line with a step where the M_x load is acting. Moment-loads are assumed to provide 'continuity'; they are the support moments in statically indeterminate, multi-span beams.

When a number of the various types of loadings act together, we may plot internal force diagrams by proceeding step by step, choosing key sections and transposing the force system of portions to the section chosen. This can be done for all the loads together, or, alternatively, typical loads can be taken separately and using the principle of superposition, we finally add together the diagrams obtained for the individual loads.

Example 5.3

A 'mixed' loading is shown in *Figure 5.8(a)* on a cantilever-type plane beam. R_y, R_z and M_x diagrams are to be plotted. It will be noted that the load at H is virtually an M_x type load, with a moment value of $1,000 \times 3$ ft $= 3,000$ lb-ft.

To find reactions, take moments about B:

$$M_{Bx} = -1,350 \times 21 - 600 \times 13 - 566 \times 6 + 1,000 \times 3 + \\ + F_{Ay} \times 20 = 0$$
$$F_{Ay} = +1,828 \text{ lb (up)}$$

Moments about A:

$$M_{Ax} = -1,350 \times 1 - 600 \times 7 + 566 \times 14 + \\ + 1,000 \times 3 - F_B \times 20 = 0$$
and $\quad F_B = +688$ lb (up)

Check: $-1,350 - 600 - 566 + 1,828 + 688 = 0$. Correct.
$$F_{Ax} = -(-1,000 + 566) = +434 \text{ lb}$$

In the following, internal forces will be calculated at key sections and plotted in *Figure 5.8(b)*, *(c)* and *(d)*.

154

Shearing force R_y: From C to A the line is a parabola, with a value just to the left of A:

$$R_{Ay} = -\frac{7}{9} \times 300 \times \frac{7}{2} = -820 \text{ lb}$$

The line continues to the right of A with a parabolic segment till D.

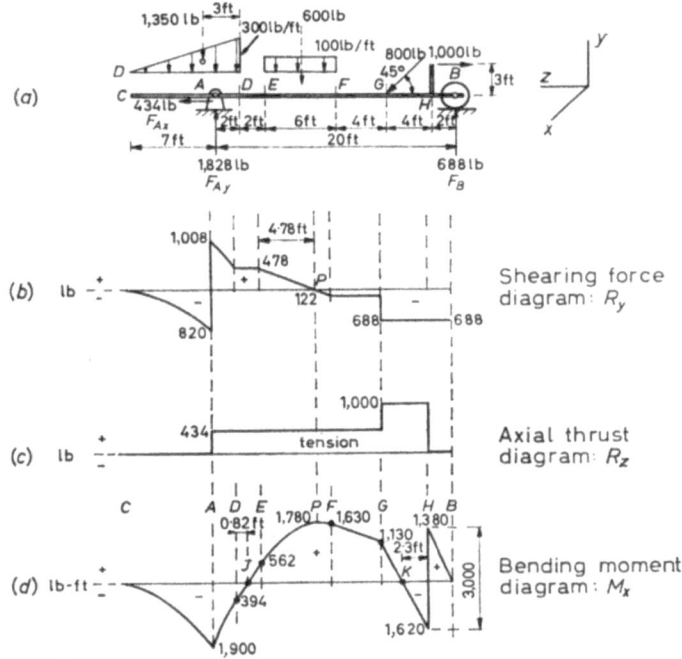

Figure 5.8. Example 5.3

The rest of the shearing force diagram can be constructed graphically.

Axial thrust R_z: $R_{Az} = +434$ lb which means tension. This remains constant to section G.

$$R_{Gz} = +434 + 566 = +1,000 \text{ lb tension}$$

Passing through H, we arrive on the stretch HB on which

$$R_z = +434 + 566 - 1,000 = 0$$

Bending moment M_x: The line on CA will be a cubic, with

$$M_{Ax} = -820 \times \frac{7}{3} = -1,900 \text{ lb-ft}$$

At D, $M_{Dx} = -1,350 \times 3 + 1,828 \times 2 = -394$ lb-ft
At E, $M_{Ex} = -1,350 \times 5 + 1,828 \times 4 = +562$ lb-ft
At F, $M_{Fx} = -1,350 \times 11 + 1,828 \times 10 - 600 \times 3 =$
$= +1,630$ lb-ft

The maximum positive M_x occurs between E and F at point P
which is $\frac{478}{100} = 4\cdot78$ ft to the right of E.

At P, $M_{Px}(\text{max}) = -1,350 \times 9\cdot78 + 1,828 \times 8\cdot78 -$
$$- 478 \times \frac{4\cdot78}{2} = +1,780 \text{ lb-ft}$$

At G, M_{Gx} $= -1,350 \times 15 + 1,828 \times 14 - 600 \times 7 =$
$= +1,130$ lb-ft

To the left of H: $M_{Hx} = -1,350 \times 19 + 1,828 \times 18 -$
$- 600 \times 11 - 566 \times 4 = -1,620$ lb-ft

To the right of H: $M_{Hx} = -1,350 \times 19 + 1,828 \times 18 -$
$- 600 \times 11 - 566 \times 4 +$
$+ 1,000 \times 3 = +1,380$ lb-ft

And, for a check, at B:

$M_{Bx} = -1,350 \times 21 + 1,828 \times 20 - 600 \times 13 - 566 \times 6 +$
$+ 1,000 \times 3 = 0$

In this case, the position of the points of contraflexure J and K
can be found from the rules of proportionality:

$$\frac{z_J}{394} = \frac{2 \text{ ft}}{394 + 562}$$

and
$$z_J = 0\cdot82 \text{ ft from } D.$$

Similarly,
$$\frac{z_K}{1,620} = \frac{4}{1,620 + 1,130}$$

and $z_K = 2\cdot3$ ft from H.

This completes Example 5.3.

Example 5.4

The simplified form of a power transmission shaft is shown in *Figure 5.9(a)*. Power is taken off, through pulleys and belts, at points C and D. At E an eccentric axial load is acting. In addition, the weight of the shaft must also be considered, at the rate of 10 lb/ft. Our task is to draw out all internal force diagrams.

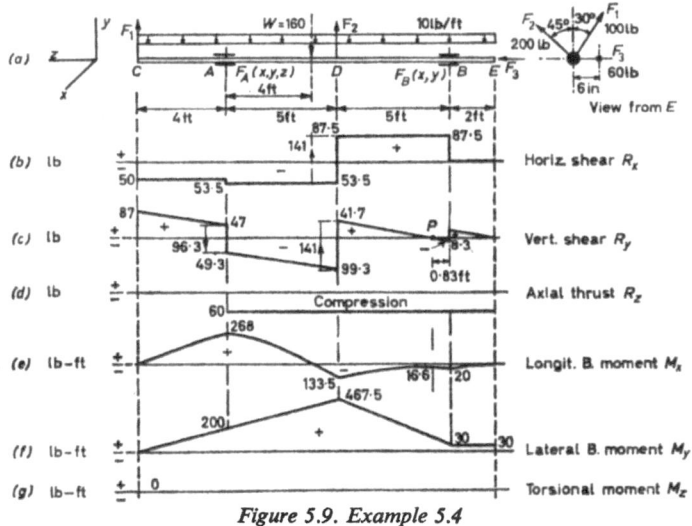

Figure 5.9. Example 5.4

From Table 5.5, forming the transposition totals and adding F_{Ax}, F_{Ay} and F_{Az}:

$$R_x = +91 + F_{Bx} + F_{Ax} = 0$$
$$R_y = +68 + F_{By} + F_{Ay} = 0$$
$$R_z = +60 + F_{Az} = 0$$
$$M_x = +283 - 10F_{By} = 0$$
$$M_y = -30 + 905 + 10F_{Bx} = 0$$
$$M_z = 0$$

From M_x: $F_{By} = +28 \cdot 3$ lb (up)
From R_y: $F_{Ay} = -96 \cdot 3$ lb (down)
From M_y: $F_{Bx} = -87 \cdot 5$ lb
From R_x: $F_{Ax} = -3 \cdot 5$ lb
From R_z: $F_{Az} = -60$ lb

157

Table 5.5. Transposition to support A

	F_x	F_y	F_z	x	y	z	$F_z.y$	$F_x.z$	$F_y.x$	$F_y.z$	$F_z.x$	$F_z.y$
	lb			ft			lb-ft					
F_1	-50	$+87$	0	0	0	$+4$	0	-200	0	$+348$	0	0
F_2	$+141$	$+141$	0	0	0	-5	0	-705	0	-705	0	0
F_3	0	0	$+60$	-0.5	0	-12	0	0	0	0	-30	0
W	0	-160	0	0	0	-4	0	0	0	$+640$	0	0
F_{Bx}	$+F_{Bx}$	0	0	0	0	-10	0	$-10F_{Bx}$	0	0	0	0
F_{By}	0	$+F_{By}$	0	0	0	-10	0	0	0	$-10F_{By}$	0	0
Totals	$+91$ $+F_{Bx}$	$+68$ $+F_{By}$	$+60$				0	-905 $-10F_{Bx}$	0	$+283$ $-10F_{By}$	-30	0

The plotting of the R_x, R_y and R_z is carried out by mechanically following and plotting loads as they occur.

M_x is plotted by considering F_y components and taking a progressive sum of moments along the beam, commencing at C.

Figure 5.10 shows F_x components along the beam. The lateral bending moment M_y is calculated progressively along the beam and plotted in the diagram. The load F_3 also affects M_y values.

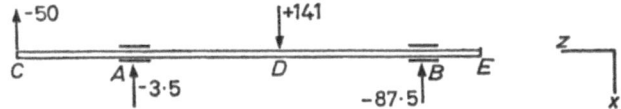

Figure 5.10. F_x components of loads and reactions, Example 5.4

This completes Example 5.4.

5.6. INTERNAL FORCES IN FRAMES

The rules for variation of internal forces and for the method of plotting them are still valid for structural elements which are vertical or on the skew or where such elements are combined with beams.

In a portal frame, two vertical columns support roof elements called 'rafters'. Such a frame is shown in *Figure 5.11* and it may be statically determinate or indeterminate, according to the type of its support restraints. A and B may be fully encased, hinged or on rollers.

Internal force diagrams may be plotted after the reactions were determined (by elastic analysis if the frame is statically indeterminate). But no matter how the reactions were calculated, once they are known, internal forces will be determined with relative ease throughout the structure.

Example 5.5

The portal frame of *Figure 5.11* is a three-hinged plane structure, thus it is statically determinate. We shall plot the three internal force diagrams for R_y, R_z and M_x.

159

Calculation of reactions at A and B:

Since the frame as a whole is in equilibrium, $M_{Bx} = 0$

$$M_{Bx} = 2{,}550 \times 7 + 30 \times 17 - 2{,}950 \times 45 - 1{,}750 \times 15 +$$
$$+ 210 \times 17 + 490 \times 7 + F_{Az} \times 60 = 0$$

and

$$F_{Az} = 2{,}230 \text{ lb}$$

Similarly, the condition $M_{Ax} = 0$ will give

$$F_{Bz} = 2{,}470 \text{ lb}$$

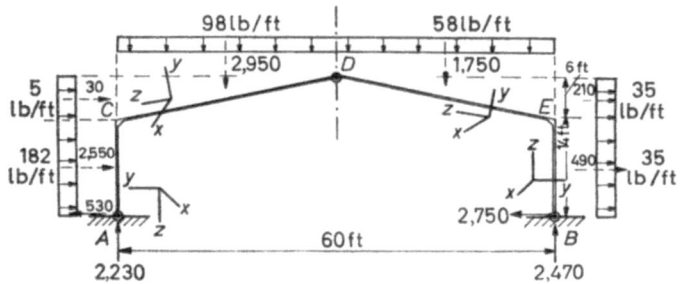

Figure 5.11. Three-hinged portal frame of Example 5.5

Check up through $R_z = 0$:

$$R_z = 2{,}230 + 2{,}470 \text{ (both up)} - 2{,}950 - 1{,}750 \text{ (both down)} = 0$$

The *horizontal reactions* cannot be found in this way. We shall write down the condition that the *bending moment at hinge D is zero*:

Bending moment at D—considering *portion AD*:

$$M_{Dx} = 2{,}230 \times 30 + F_{Ay} \times 20 - 2{,}550 \times 13 - 30 \times 3 -$$
$$- 2{,}950 \times 15 = 0$$

and $F_{Ay} = 530$ lb (as assumed, right to left).

Bending moment at D—considering *portion BD*:

$$M_{Dx} = 2{,}470 \times 30 - F_{By} \times 20 + 490 \times 13 + 210 \times 3 -$$
$$- 1{,}750 \times 15 = 0$$

that is

$$F_{By} = 2{,}750 \text{ lb (right to left, as assumed).}$$

160

Check up through $R_y = 0$:

$$R_y = +530 + 2{,}750 - 2{,}550 - 30 - 210 - 490 = 0.$$

This completes finding the four reaction components at A and B. Internal force variations are shown in the following diagrams.

Shearing force and axial thrust

R_y is shown in *Figure 5.12*. This is now the shearing force component of the internal force, *everywhere at right angles* to the line of the structure, since the cross section, the plane of the shearing force,

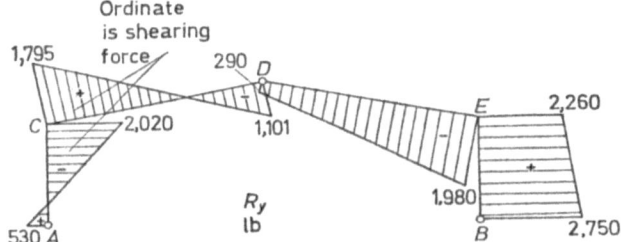

Figure 5.12. Shearing force diagram of Example 5.5

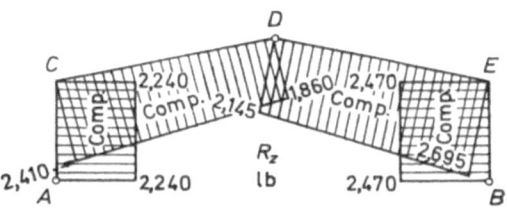

Figure 5.13. Axial thrust diagram of Example 5.5

is at right angles to the centroid line. The direction of R_y therefore will vary—from A to C it will be horizontal, from C to D it will be at right angles to the line CD and so on.

Note that values of R_y were plotted on the *centroid line* which was

161

used as datum. A similar diagram was plotted for axial thrust, R_z, in *Figure 5.13.*

The plotting of R_y and R_z along AC is straightforward. At A the reaction components F_{Ay} (*Figure 5.12*) and F_{Az} (*Figure 5.13*) are plotted at right angles to the datum (centroid line). The shearing force will vary from A to C by 182 lb/ft $\times$ 14 ft = 2,550 lb opposite to F_{Ay} = 530 lb. That is, net shear just below C is 2,550 − 530 = 2,020 lb left to right. Axial thrust is constant and *compression* from A to C.

The two horizontal and vertical components just below C will now have to be converted into a component at right angles to CD (shearing force) and parallel to CD (axial thrust) as we move along the rafter CD, towards D.

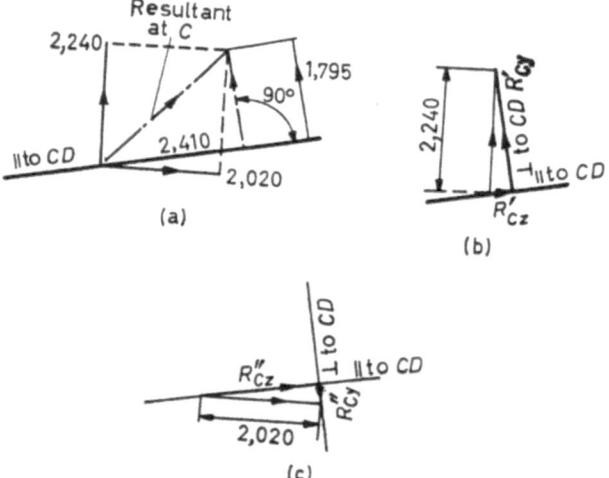

Figure 5.14. Shearing force and axial thrust along the sloping member CD

This conversion for CD can be carried out graphically as shown in *Figure 5.14(a)* where the resultant at C is resolved into the shear and thrust directions of member CD. *Figures 5.14(b)* and *(c)* illustrate a suitable way of calculation, using proportionalities, involving the length of CD:

$$L_{OD} = \sqrt{6^2 + 30^2} = 30 \cdot 6 \text{ ft}$$

Then $\quad \dfrac{30 \cdot 6}{30} = \dfrac{2,240}{R_{Oy}'}$ and $R_{Oy}' = +2,190 \text{ lb}$

See Figure 5.14(b)

Also $\quad \dfrac{30 \cdot 6}{6} = \dfrac{2,240}{R_{Oz}'}; \quad R_{Oz}' = -440 \text{ lb}$

From *Figure 5.14(c)*:

$$\frac{30 \cdot 6}{6} = \frac{2,020}{R_{Oy}''}; \quad R_{Oy}'' = -395 \text{ lb}$$

$$\frac{30 \cdot 6}{30} = \frac{2,020}{R_{Oz}''}; \quad R_{Oz}'' = -1,970 \text{ lb}$$

Summing up:

$R_{Oy} = R_{Oy}' + R_{Oy}'' = +2,190 - 395 = +1,795 \text{ lb}$ (shearing force)
$R_{Oz} = R_{Oz}' + R_{Oz}'' = -440 - 1,970 = -2,410 \text{ lb}$ (axial thrust).

These values were plotted at *C*, on rafter *CD*, in *Figures 5.12* and *5.13*. Similarly, we obtain the shear and axial components of the *loads* on *CD*. This resolution is shown diagrammatically in *Figure 5.15*; for convenience, the vertical (2,950 lb) and horizontal (30 lb) total loads on *CD* have been resolved (*a*) parallel, and (*b*) at right angles, to *CD*.

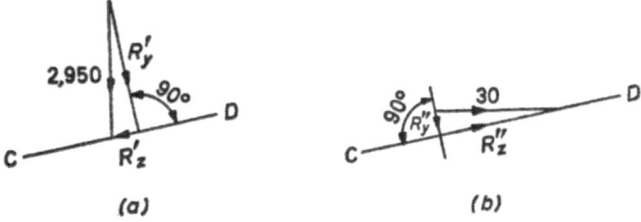

Figure 5.15. Shear and axial components of loads on CD

The same proportionality as above will be used:

$R_y' = \dfrac{30}{30 \cdot 6} \times 2,950 = -2,890 \text{ lb}, \quad R_y'' = \dfrac{6}{30 \cdot 6} \times 30 = -6 \text{ lb}$
$R_y = -2,896 \text{ lb}$

163

$$R_z{'} = \frac{6}{30 \cdot 6} \times 2{,}950 = +580 \text{ lb}, \qquad R_z{''} = \frac{30}{30 \cdot 6} \times 30 = -30 \text{ lb}$$

$$R_z = +550 \text{ lb}$$

These values represent the *total* (uniform) *change* of the shearing force and axial thrust on CD and were plotted in this way in *Figures 5.12* and *5.13*.

The DEB part of the structure can be handled in either of two ways. We can continue on to DE and passing through Section D, repeat the operation previously performed at C, needed because of a change in the direction of the centroid line. Or, we may go back to B and start plotting up from B, continuing along ED (remembering that the sign of internal forces in this plotting must be reversed).

In either case, a check is available: using the first method, our plot must finish at B with F_{By} and F_{Bz}; with the second, internal force components at, say, the left of D must be the same whether we arrived there plotting from A or from B. The complete diagrams are shown in *Figures 5.12* and *5.13* and the reader is advised to work out the DEB part independently and compare his results with the diagrams shown.

Bending moment

The bending moment, M_x, is plotted in *Figure 5.16*. For upright AC, equation of M_x can be written as

$$M_x = 530 \times z - 182 \times \frac{z^2}{2}$$

This will have *zero value* where

$$530z - 182\frac{z^2}{2} = 0$$

or

$$91z = 530$$

$$z = 5 \cdot 8 \text{ ft}$$

M_x will be *maximum* where $R_y = 0$ and from *Figure 5.12* this occurs at

$$z_N = \frac{530}{530 + 2{,}020} \times 14 \text{ ft} = 2 \cdot 9 \text{ ft}$$

Actually, the M_x line is a parabola which is symmetrical about the horizontal line through point N, 2·9 ft above A.

$$M_x \text{ at } N = 530 \times 2 \cdot 9 - 182 \times \frac{2 \cdot 9^2}{2} = +775 \text{ lb-ft}$$

$$M_x \text{ at } C = 530 \times 14 - 182 \times \frac{14^2}{2} = -10{,}400 \text{ lb-ft}$$

At C, turning on to rafter CD, M_x will remain unchanged since forces and their distances from C are the same whether we take a section just below C or just to the right of it on CD.

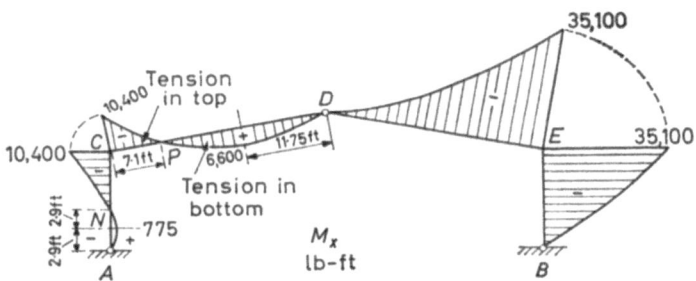

Figure 5.16. Bending moment diagram for **M**ₓ

M_x is written for CD in equation form as follows, with reference to section P:

Since $\quad z_v = z\dfrac{6}{30\cdot6} = 0\cdot195z, \quad$ and $\quad z_h = z\dfrac{30}{30\cdot6} = 0\cdot98z,$

$$M_{Px} = 530(14 + z_v) + 2{,}240 \times z_h - 2{,}550(7 + z_v) - \\ - 98 \times \frac{z_h^2}{2} - 5\frac{z_v^2}{2}$$

Substituting the above values of z_v and z_h and simplifying:

$$M_{Px} = -10{,}400 + 1{,}805z - 48z^2$$
M_x will be zero where $M_{Px} = 0$. This results in
$z = 7\cdot1$ ft or $30\cdot6$ ft

Along CD, the hinge D is at a distance of $z = 30\cdot6$ ft from C. Here M_x is obviously zero. The other value, $z = 7\cdot1$ ft gives the distance of Section P from C.

Since the axis of symmetry of the parabola will again be midway between P and D, this is the section for maximum M_x on CD.

This point is at

$$\frac{30\cdot6 - 7\cdot1}{2} = 11\cdot75 \text{ ft from } D, \text{ measured along } CD \text{ or at}$$
$$z = 7\cdot1 + 11\cdot75 = 18\cdot8 \text{ ft from } C.$$

The value of M_x here is

$$-10,400 + 1,805 \times 18\cdot8 \text{ ft} - 48 \times (18\cdot8)^2 = +6,600 \text{ lb-ft}$$

All this is shown in *Figure 5.16*, also the M_x line for *DE* and *EB* which the reader should check.

This completes Example 5.5.

5.7. INTERNAL FORCES IN TRUSSES

In this chapter we have applied the rules of equilibrium to parts of *solid bodies*. In this way, internal forces at any section of the body were determined. As we have seen, internal forces usually vary throughout the solid structure and we used force diagrams to show this variation.

Articulated bodies or trusses have loads acting at joints only and *all internal forces are axial thrust forces* (tension or compression) in the members. This means that there are no shearing forces, bending or twisting moments in trusses, only axial thrust and that is constant within the length of a given member, between its two joints.

The principle of equilibrium can still be applied to the 'solution' of trusses. Broadly, there are two methods:

(*a*) establishing the equilibrium of *a joint* under the combined action of loads acting on the joint and internal forces of members meeting at the joint, or

(*b*) assuming that a *portion* of the truss is in equilibrium under the action of loads and reaction on the portion and that the axial thrust forces in the 'section' bounding the truss-portion will hold the resultant of external forces in equilibrium.

Either method may be used in a given case. The first method, known as *equilibrium of pin* method, has been demonstrated in Chapter 4 both graphically and by calculation—*see* Example 4.10 with *Figures 4.25* to *4.28* inclusive and Example 4.13 with *Figure 4.32*. The graphical construction suits only plane structures but the tabulated calculation of Example 4.13 (in Table 4.2) can be used for both space and plane structures. In cases when the structure has several free joints, the table of forces may have to be set up for each of the joints in turn. Since there are only three equations available (x, y and z components of the forces), we can get a clear-cut solution for a joint if there are three members meeting there. When more than three members meet at a joint, the equations cannot be readily solved and would have to be 'pooled' with similar equations written out for other joints. Ultimately, if the structure is stable and determinate, the number of unknowns (axial thrusts) should equal the total number of simultaneous equations.

Even simple space structures may yield a great number of equations in this way and whenever it is possible, we should start at a joint where only three members meet, start from there and proceed to the next joint, using the previously obtained values for members in subsequent equations.

This successive, joint-by-joint approach has been used in the graphical method for plane trusses. Errors can easily be carried forward in these calculations and a check is only available at the end when we arrive at a reaction which must hold the last group of solved members in equilibrium.

The second method, that of working with a portion of the structure, is known as the *method of sections*. This resembles calculations we have used in determining internal forces in solid bodies. It involves taking a 'cut' through the truss and *restoring it by substituting internal forces* in the cut members. In this way, the internal forces can be determined. Each 'cut' is taken independently and there is no cumulative error.

In the following sections of this chapter, these methods will be demonstrated.

5.8. EQUILIBRIUM OF PIN METHOD

Example 5.6

The space truss in *Figure 5.17* has two free joints and it is just stiff since $3 \times 2 = 6$ members connect the free joints A and B to the ground supports C, D and E. At point B only three members are meeting and we will commence our tabulated calculation there.

167

Assume all members in tension as shown by arrows in Plan at B. Length of members meeting at B:

$$BA = \sqrt{6^2 + 4^2} = 7 \cdot 2 \text{ ft}$$
$$BD = \sqrt{7^2 + 16^2 + 10^2} = 20 \cdot 1 \text{ ft}$$
$$BE = 20 \cdot 1 \text{ ft}$$

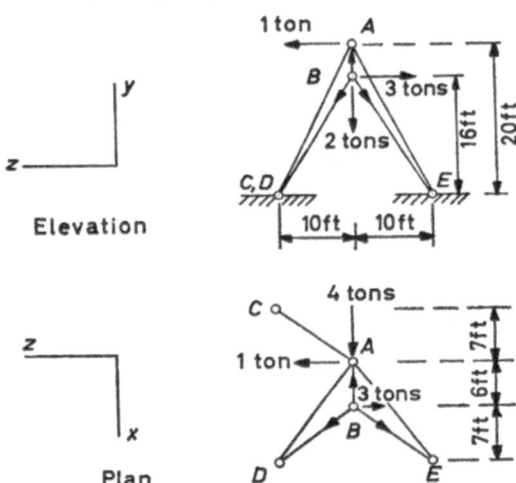

Figure 5.17. Space truss of Example 5.6

Table 5.6. Equilibrium of pin B, Example 5.6

	F_x	F_y	F_z
2 tons	0	-2	0
3 tons	0	0	-3
F_{BA}	$-\dfrac{6}{7 \cdot 2} \cdot F_{BA}$	$+\dfrac{4}{7 \cdot 2} \cdot F_{BA}$	0
F_{BD}	$+\dfrac{7}{20 \cdot 1} \cdot F_{BD}$	$-\dfrac{16}{20 \cdot 1} \cdot F_{BD}$	$+\dfrac{10}{20 \cdot 1} \cdot F_{BD}$
F_{BE}	$+\dfrac{7}{20 \cdot 1} \cdot F_{BE}$	$-\dfrac{16}{20 \cdot 1} \cdot F_{BE}$	$-\dfrac{10}{20 \cdot 1} \cdot F_{BE}$

Totals: see below.

$$x: -\frac{6}{7 \cdot 2} \cdot F_{BA} + \frac{7}{20 \cdot 1} \cdot F_{BD} + \frac{7}{20 \cdot 1} \cdot F_{BE} = 0$$

$$y: -2 + \frac{4}{7 \cdot 2} \cdot F_{BA} - \frac{16}{20 \cdot 1} \cdot F_{BD} - \frac{16}{20 \cdot 1} \cdot F_{BE} = 0$$

$$z: -3 + \frac{10}{20 \cdot 1} \cdot F_{BD} - \frac{10}{20 \cdot 1} \cdot F_{BE} = 0$$

Rewriting, using decimals:

$$x: -8 \cdot 4 F_{BA} + 3 \cdot 5 F_{BD} + 3 \cdot 5 F_{BE} \qquad = 0$$
$$y: \quad 5 \cdot 5 F_{BA} - 7 \cdot 9 F_{BD} - 7 \cdot 9 F_{BE} - 20 = 0$$
$$z: \qquad\qquad 5 F_{BD} - 5 F_{BE} - 30 \quad = 0$$

From z: $\qquad\qquad F_{BE} = F_{BD} - \dfrac{30}{5} = F_{BD} - 6$

Substituting into x and y:

$$x: -8 \cdot 4 F_{BA} + 3 \cdot 5 F_{BD} + 3 \cdot 5 F_{BD} - 21 \qquad = 0$$
$$y: \quad 5 \cdot 5 F_{BA} - 7 \cdot 9 F_{BD} - 7 \cdot 9 F_{BD} + 47 \cdot 4 - 20 = 0$$

that is,

$$x: -8 \cdot 4 F_{BA} + 7 F_{BD} - 21 \qquad = 0$$
$$y: \quad 5 \cdot 5 F_{BA} - 15 \cdot 8 F_{BD} + 27 \cdot 4 = 0$$

From x: $F_{BD} = 3 + 1 \cdot 2 F_{BA}$, and substituting into y:

$$5 \cdot 5 F_{BA} - 47 \cdot 4 - 19 F_{BA} + 27 \cdot 4 = 0$$
$$F_{BA} = -\frac{20}{13 \cdot 5} = -1 \cdot 48 \text{ tons (compression)}$$
$$F_{BD} = 3 - 1 \cdot 2 \times 1 \cdot 48 = +1 \cdot 22 \text{ tons (tension, as assumed)}$$
$$F_{BE} = 1 \cdot 22 - 6 = -4 \cdot 78 \text{ tons (compression)}.$$

Two of the assumed signs in *Figure 5.17* were thus proved incorrect. In *Figure 5.18* the correct signs are shown at B, together with *assumed* signs at A for AC, AD and AE (all assumed in tension). In Table 5.7 the known value of AB will be used (now known to be in compression), thus leaving three unknown forces to be found.

Length of members meeting at A:

$$AB = 7 \cdot 2 \text{ ft}$$
$$AC = \sqrt{7^2 + 20^2 + 10^2} = 23 \cdot 4 \text{ ft}$$
$$AD = \sqrt{13^2 + 20^2 + 10^2} = 25 \cdot 9 \text{ ft}$$
$$AE = 25 \cdot 9 \text{ ft}$$

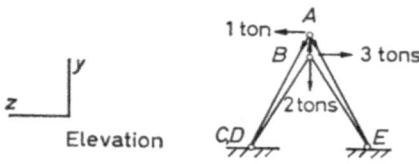

Elevation

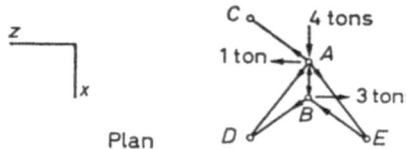

Plan

Figure 5.18. Example 5.6; Correct signs of members at **B**; *assumed signs at* **A**

Table 5.7. Equilibrium of pin A, Example 5.6.

	F_x	F_y	F_z
1 ton	0	0	$+1$
4 tons	$+4$	0	0
F_{AB}	$-\dfrac{6}{7\cdot2}F_{AB}$	$+\dfrac{4}{7\cdot2}F_{AB}$	0
F_{AC}	$-\dfrac{7}{23\cdot4}F_{AC}$	$-\dfrac{20}{23\cdot4}F_{AC}$	$+\dfrac{10}{23\cdot4}F_{AC}$
F_{AD}	$+\dfrac{13}{25\cdot9}F_{AD}$	$-\dfrac{20}{25\cdot9}F_{AD}$	$+\dfrac{10}{25\cdot9}F_{AD}$
F_{AE}	$+\dfrac{13}{25\cdot9}F_{AE}$	$-\dfrac{20}{25\cdot9}F_{AE}$	$-\dfrac{10}{25\cdot9}F_{AE}$

Substituting $1\cdot48$ tons for F_{AB} which is acting in the direction shown in *Figure 5.18*,

$$x: +27\cdot7 - 3F_{AC} + 5F_{AD} + 5\,F_{AE} = 0$$
$$y: +8\cdot3 - 8\cdot5F_{AC} - 7\cdot7F_{AD} - 7\cdot7F_{AE} = 0$$
$$z: +10 + 4\cdot25F_{AC} + 3\cdot85F_{AD} - 3\cdot85F_{AE} = 0$$
$$2z: +20 + 8\cdot5F_{AC} + 7\cdot7F_{AD} - 7\cdot7F_{AE} = 0$$
$$y + 2z: +28\cdot3 \qquad\qquad\qquad - 15\cdot4F_{AE} = 0$$

$$F_{AE} = +1\cdot83 \text{ tons (tension)}$$

170

Substituting this value into x and y, we get

$$F_{AD} = -5\cdot05 \text{ tons (compression)}$$
$$F_{AC} = +3\cdot87 \text{ tons (tension)}$$

Note that the arrow on AD in *Figure 5.18* must now be reversed to show AD correctly in compression.

This completes Example 5.6.

The equilibrium of pin method can be summed up as follows:

(*a*) Select a joint, preferably one which has only three members meeting.

(*b*) Calculate the true length of each member from the formula $L = \sqrt{x^2 + y^2 + z^2}$ where x, y, z are the co-ordinates of each member concerned.

(*c*) Assume that all members at the joint are in tension.

(*d*) Form a tabulation of the F_x, F_y and F_z components of *all* forces at the joint. Form the F_x component by using the cosine factor, that is, x component of length divided by L. F_y and F_z are formed similarly.

(*e*) The sum total of each of F_x, F_y and F_z must be zero for the equilibrium of the joint. This yields three equations for the three unknown members meeting at the joint.

(*f*) Solve the equations: if the sign is $+$, the assumed tension is correct. If the sign is $-$, the member is in compression.

(*g*) If there is no joint with only three members meeting, several tabulations must be set up and the equations will then be solved simultaneously.

For a *plane truss*, two equations of equilibrium will be available at every joint. This means that if we can commence our calculation at a joint where only two members are meeting, the two unknown forces will be obtained and in this manner we may proceed from joint to joint. This has been carried out in Example 4.10 by graphical means; it will now be demonstrated by calculation.

Example 5.7

In *Figure 5.19*, a plane truss is loaded at joints ('panel points'). What are the forces in its members?

171

Geometry: From the dimensions it follows that *DH* and *FJ* are vertical and angle *CHA* = angle *CAH*.

Vertical component of 800 lb load: $800 \times \dfrac{10}{11 \cdot 2} = 716$ lb

Horizontal component of 800 lb load: $800 \times \dfrac{5}{11 \cdot 2} = 358$ lb

$HE = EJ = \sqrt{5^2 + 3 \cdot 33^2} = 6$ ft

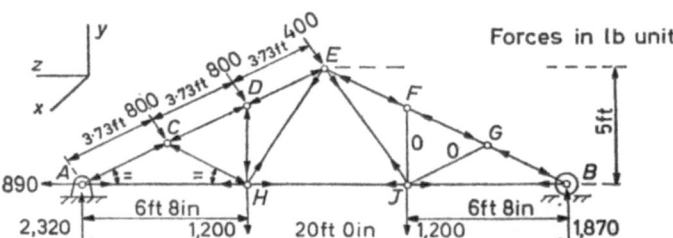

Figure 5.19. Truss of Example 5.7

Reactions: By taking moments M_x about first *A*, then *B*, we find F_B, F_{4y} and finally also F_{4z} which is equal and opposite to the total horizontal component of the acting loads. We remember to include both vertical and horizontal components of the acting loads when taking moments.

$$F_{By} = +1{,}870 \text{ lb (up)}$$
$$F_{Ay} = +2{,}320 \text{ lb (up)}$$
$$F_{Az} = +\ \ 890 \text{ lb (from right to left)}.$$

Equilibrium of pins: We commence at *A* and from here proceed to *C*, then to *D*, to *H*, to *E* and across to the right half of the truss. The calculations are tabulated in Table 5.8. All unknown members are initially assumed to be in tension.

Finally, a check can be made at *B*:

y: $+1{,}870 + \dfrac{5}{11 \cdot 2}F_{GB} = 0$ and $F_{GB} = -4{,}200$ lb (compression)

z: $-\dfrac{10}{11 \cdot 2} \times 4{,}200 + F_{JB} = 0$, that is $F_{JB} = +3{,}730$ lb (tension).

These check up with the values obtained in Table 5.8.

172

Joint	Component	Equation	Force lb	
A	y	$+2{,}320 + \dfrac{5}{11\cdot 2}F_{AO} = 0$	F_{AO} 5,200	c
	z	$+890 - \dfrac{10}{11\cdot 2}(-5{,}200) - F_{AH} = 0$	F_{AH} 5,530	t
C	y	$-716 + 2{,}320 + \dfrac{5}{11\cdot 2}F_{CD} - \dfrac{5}{11\cdot 2}F_{CH} = 0$	F_{CD} 4,590	c
	z	$-358 - \dfrac{10}{11\cdot 2}5{,}200 - \dfrac{10}{11\cdot 2}F_{CD} - \dfrac{10}{11\cdot 2}F_{CH} = 0$	F_{CH} 990	c
D	y	$-716 + \dfrac{5}{11\cdot 2}4{,}590 - F_{DH} + \dfrac{5}{11\cdot 2}F_{DE} = 0$	F_{DH} 880	c
	z	$-358 - \dfrac{10}{11\cdot 2}4{,}590 - \dfrac{10}{11\cdot 2}F_{DE} = 0$	F_{DE} 4,960	c
H	y	$\dfrac{5}{11\cdot 2}990 - 880 + \dfrac{5}{6}F_{HE} - 1{,}200 = 0$	F_{HE} 3,020	t
	z	$+5{,}530 - \dfrac{10}{11\cdot 2}990 - \dfrac{3\cdot 33}{6}\times 3{,}020 - F_{HJ} = 0$	F_{HJ} 2,940	t
E	y	$-358 + \dfrac{5}{11\cdot 2}4{,}960 - \dfrac{5}{6}3{,}020 - \dfrac{5}{6}F_{EJ} - \dfrac{5}{11\cdot 2}F_{EF} = 0$	F_{EJ} 1,440	t
	z	$-179 - \dfrac{10}{11\cdot 2}\times 4{,}960 + \dfrac{3\cdot 33}{6}3{,}020 - \dfrac{3\cdot 33}{6}F_{EJ} - \dfrac{10}{11\cdot 2}F_{EF} = 0$	F_{EF} 4,200	c
F	y, z	By inspection $F_{GE} =$	F_{EG} 4,200	c
J	y	$+\dfrac{5}{6}1{,}440 - 1{,}220 + F_{FJ} = 0$	F_{FJ} 0	—
	z	$+2{,}940 + \dfrac{3\cdot 33}{6}1{,}440 - F_{JB} = 0$	F_{JB} 3,730	t

Apart from the possibility of errors, the method is laborious. For comparison, a graphical construction ('stress diagram') was prepared in *Figure 5.20* which demonstrates the simplicity of the graphical approach.

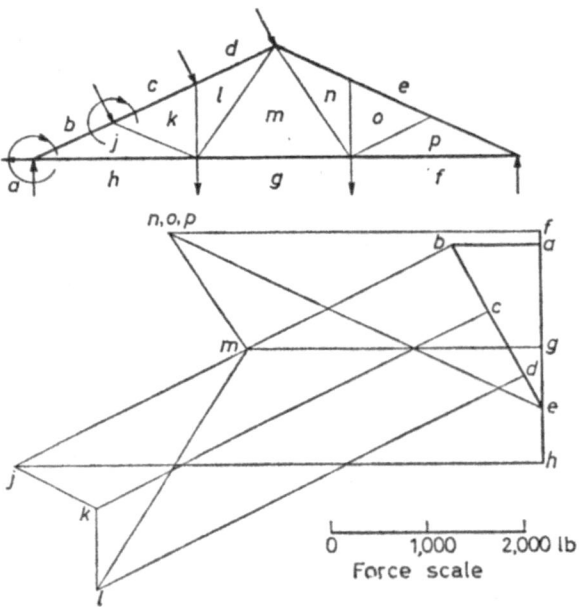

Figure 5.20. Stress diagram for Example 5.7

In *Figure 5.20* the so-called 'Bow's notation' was used and points of intersection in the force diagram correspond to spaces between members in the space diagram. External forces (including reactions) were plotted in a *clockwise direction* around the structure. Construction started at space *h* and members were plotted in a *clockwise sequence h—a—b—j—h*. This gave a closed polygon in the force diagram, since at joint *A forces are in equilibrium*. Construction proceeded to joint *C* and here the sequence was *j—b—c—k—j*. Incidentally, signs (tension or compression) are determined by

174

following the sequence around in the force polygon and transferring the signs to the joint concerned.

Forces can be read off the diagram at the adopted force scale: for instance, force $d - l$ is F_{DE} and force $l - k$ is F_{DH}.

This concludes Example 5.7.

5.9. THE METHOD OF SECTIONS

The method will be illustrated on the *plane* truss of Example 5.7 which is reproduced in *Figure 5.21*.

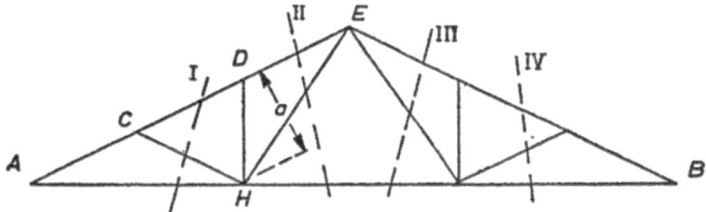

Figure 5.21. Sections through truss of Example 5.7

We may now 'cut' the truss, obtain a portion and examine it on the principle of equilibrium of a part. Let us make cut I in *Figure 5.21* and re-draw portion *A*—I in *Figure 5.22(a)* with the loads acting on it.

Our task is to find three forces F_{CD}, F_{CH} and F_{AH} acting along these members, *such that the loads are held in equilibrium.*

Since we have three conditions for the equilibrium of a plane body, we should be able to find three unknown forces. Applying this principle in general, we would obtain three simultaneous equations:

$$R_y = 0$$
$$R_z = 0$$
$$M_x = 0 \text{ (about any point in the plane).}$$

The work of solving simultaneous equations can be saved by working step by step as follows:

175

If the point for M_x is taken at C, two of the unknown forces, F_{CD} and F_{CH} meet here and these will therefore not contribute to M_x. The equation of M_x will contain only one unknown, F_{AH}:

$$M_{Cx} = +2{,}320 \times 3{\cdot}33 + 890\frac{3{\cdot}33}{2} - F_{AH} \times \frac{3{\cdot}33}{2} = 0$$

and

$$F_{AH} = +5{,}530 \text{ lb (tension, as assumed).}$$

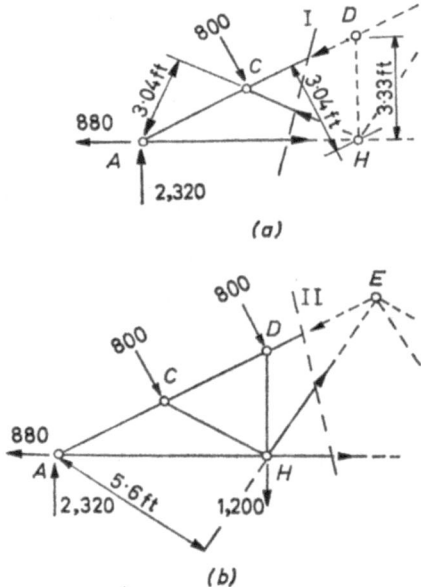

(a)

(b)

Figure 5.22. Portions A—I and A—II of Example 5.7

In this, C was the 'moment point' and the same operation can be repeated for moment points H and A:

176

$$M_{Hx} = +2{,}320 \times 6 \cdot 67 - 716 \times 3 \cdot 33 + 358 \times \frac{3 \cdot 33}{2} -$$
$$- F_{CD} \times 3 \cdot 04 = 0$$

$F_{CD} = +4{,}590$ lb (compression, as assumed).

$$M_{Ax} = +716 \times 3 \cdot 33 + 358 \times \frac{3 \cdot 33}{2} - F_{CH} \times 3 \cdot 04 = 0$$

$F_{CH} = +990$ lb (compression, as assumed).

Note that signs on the unknown members were assumed and moments entered accordingly. If the result is positive, this means that the assumed direction was correct. The moment arm 3·04 ft was obtained from simple proportionality.

Only *loads on the portion* considered are included; moment points, however, may be anywhere on or outside the structure. It does not matter exactly where cut I is taken as long as it slices through the three members considered.

Next, we take cut II and show portion A—II in *Figure 5.22(b)* with its loads.

Moment point H:

$$M_{Hx} = +2{,}320 \times 6 \cdot 67 - 716 \times 3 \cdot 33 + 358 \times \frac{3 \cdot 33}{2} +$$
$$+ 358 \times 3 \cdot 33 - F_{DE} \times 3 \cdot 04 = 0$$

$F_{DE} = +4{,}960$ lb (compression)

Moment point E:

$$M_{Ex} = +2{,}320 \times 10 + 890 \times 5 - 1{,}600 \times \frac{11 \cdot 2}{2} -$$
$$- 1{,}200 \times 3 \cdot 33 - F_{HJ} \times 5 = 0$$

$F_{HJ} = +2{,}940$ lb (tension).

Moment point A:

$$M_{Ax} = +1{,}600 \times \frac{11 \cdot 2}{2} + 1{,}200 \times 6 \cdot 67 - F_{HE} \times 5 \cdot 6 = 0$$

$F_{HE} = +3{,}020$ lb (tension).

The calculation to find the moment arm for *HE* may be

cumbersome. In such cases, we may use the $R_y = 0$ condition for portion A—II.

$$R_{\mathrm{II}y} = +2{,}320 - 2 \times 716 - 1{,}200 -$$
$$- \frac{5}{11 \cdot 2} \times F_{DE} + \frac{5}{6} F_{HE} = 0$$

and, substituting $F_{DE} = 4{,}960$,

$$F_{HE} = 3{,}020 \text{ lb}$$

The $R_y = 0$ condition is mainly useful when web members in parallel chord trusses are calculated.

Naturally, we can check axial thrust results obtained by the Method of Sections with the graphical construction and pin-by-pin calculation of *Figure 5.20* and Table 5.8. The student is urged to complete the problem by taking Sections III and IV in *Figure 5.21*. Portions may be assumed from *A or B*; results should be checked with Table 5.8.

The Method of Sections may also be used to solve *space trusses*. Basic principles are the same; now we have six conditions (equations) and in order to get direct results, we should endeavour to slice through not more than six members.

Example 5.8

The articulated tower shown in *Figure 5.23* is just stiff: starting from ground supports A, B and C each new triangular horizontal set of *three* joints is held to the base structure by *nine* new members.

Vertical loads: 200 lb on *every* joint, down.
Horizontal loads: 50 lb $-x$ direction, on joints A
                            50 lb $-z$ direction, on joints A
                            50 lb $-x$ direction, on joints C

We take cut I just above the ground supports. Six members are cut:

$$A_1 A_0, \quad B_1 B_0, \quad C_1 C_0$$
$$B_1 A_0, \quad C_1 B_0, \quad A_1 C_0$$

All are assumed to be in tension. Next, we transpose the entire load system above cut I (that is, we consider portion I—top) to point A_0. This is carried out in Table 5.9. The first three lines stand for the external loads: acting vertically; horizontally on A and C in the $-x$

178

direction; and horizontally on A in the $-z$ direction. For all three types of load, their *resultant* for the portion *above* cut I was entered. The other six lines represent the six forces sliced by the cut.

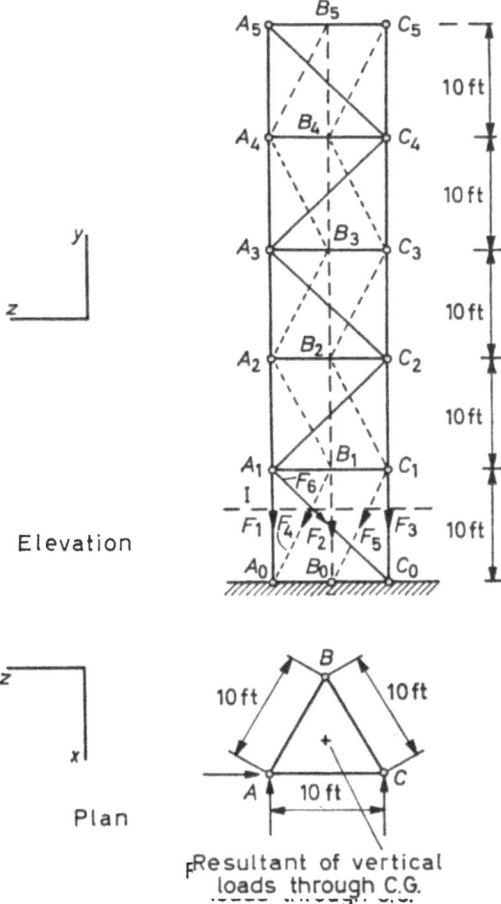

Elevation

Plan

Resultant of vertical loads through C.G.

Figure 5.23. Space truss of Example 5.8

179

Table 5.9. Example 5.8. Transposition to A_0, Section I

	F_x	F_y	F_z	x	y	z	$F_z.y$	$F_x.z$	$F_y.x$	$F_y.z$	$F_z.x$	$F_z.y$
	lb			ft			lb-ft					
Vert.	0	−3,000	0	−2.9	0	−5	0	0	+8,700	+15,000	0	0
Horiz. A,C	−500	0	0	0	+30	−5	−15,000	+2,500	0	0	0	0
Horiz. A	0	0	−250	0	+30	0	0	0	0	0	0	−7,500
F_1	0	$-F_1$	0	0	0	0	0	0	0	0	0	0
F_2	0	$-F_2$	0	−8.6	0	−5	0	0	+8.6F_2	+5F_2	0	0
F_3	0	$-F_3$	0	0	0	−10	0	0	0	+10F_3	0	0
F_4	+0.61F_4	−0.7F_4	+0.35F_4	0	0	0	0	0	0	0	0	0
F_5	−0.61F_5	−0.7F_5	+0.35F_5	−8.6	0	−5	0	+3.1F_5	+6F_5	+3.5F_5	−3F_5	0
F_6	0	−0.7F_6	−0.7F_6	0	0	−10	0	0	0	+7F_6	0	0

True lengths of skew members:

$$L = \sqrt{8\cdot65^2 + 10^2 + 5^2} = 14\cdot3 \text{ ft}$$

R_x: $\quad -500 + 0\cdot61F_4 - 0\cdot61F_5 = 0$

R_y: $\quad -3,000 - F_1 - F_2 - F_3 - 0\cdot7F_4 - 0\cdot7F_5 - 0\cdot7F_6 = 0$

R_z: $\quad -250 + 0\cdot35F_4 + 0\cdot35F_5 - 0\cdot7F_6 = 0$

M_x: $\quad +15,000 + 5F_2 + 10F_3 + 3\cdot5F_5 + 7F_6 - (-7,500) = 0$

M_y: $\quad -3F_5 - 2,500 - 3\cdot1F_5 = 0$

M_z: $\quad -15,000 - (8,700 + 8\cdot6F_2 + 6F_5) = 0$

From M_y: $F_5 = -\dfrac{2,500}{6\cdot1} \quad = -410$ lb (compression)

Into M_z: $F_2 = -\dfrac{26,140}{8\cdot6} = -3,040$ lb (compression)

Into R_x: $F_4 = +\dfrac{250}{0\cdot61} \quad = +410$ lb (tension)

Into R_z: $F_6 = -\dfrac{250}{0\cdot7} \quad = -355$ lb (compression)

Into M_x: $F_3 = -\dfrac{3,380}{10} \quad = -338$ lb (compression)

Into R_y: $F_1 \quad\quad\quad = +628$ lb (tension)

This gives the values of the six members cut by Section I. Other sections can be taken, along upper bays of the tower and all forces in vertical planes can be thus found. The method does not work with members in horizontal frames; such as $A_1 B_1$ etc. The equilibrium of pin method may then be applied to joint A_1. Note, however, that $F_{A_0 A_1}$ and $F_{A_1 A_2}$ will not influence the value of $A_1 B_1$ and these two forces may be omitted from the calculation at pin A_1.

The student is advised to work through at least another section of the tower and one equilibrium of pin problem, say, at a joint A.

Incidentally, Table 5.9 will lead us direct to the values of reactions at A_0, B_0 and C_0. The reaction at A_0 is equal and opposite to the resultant of F_1 and F_4 and similarly for the other supports.

Summarizing the Method of Sections for the solution of a space truss:

(a) Cut structure at a section where possibly only six members are sliced. Regard one of the separated portions of the structure as being in equilibrium under the combined effect of the acting loads and the forces in the severed members. Note that the acting loads must include *reactions on the portions*; in the case of a cantilever, it is advantageous therefore to select a portion towards the free end of the cantilever—this will not include the supports and the reactions there.

(*b*) Carry out transposition to a convenient point. This point may be *anywhere* in space; the forces of Example 5.8 were transposed to a point where two forces met, this led to simpler calculations.

(*c*) Assume tension in all cut members, this means that the forces pull away from the portion. F_x, F_y, F_z in the transposition will have signs accordingly. For x, y, z again choose a convenient point on the *line of each force*. Naturally, correct signs are most important in all this.

5.10. SUMMARY: EQUILIBRIUM OF PARTS

The principle of equilibrium of parts is one of the most powerful tools of Statics. It allows us to delve into the structure and analyse its internal forces.

For solid, articulated or framed structures in equilibrium the method of sections will yield a 'portion' of the structure which in itself is in equilibrium. From this starting assumption the restoring forces in the section, the internal forces, will be determined.

In articulated structures, we may isolate one of the joints and examine it as a 'part' in equilibrium under the combined action of loads acting on it and the forces in members meeting there. This method can be used graphically or by calculation.

Whenever we wish to establish equilibrium: for finding reactions or internal forces, transposition offers a systematic method.

5.11. PROBLEMS

Problem 5.1

The plane frame *ACEB* is hinged at *A* and supported on plane rollers at *B* as shown in *Figure 5.24*. Draw out labelled shearing force, axial thrust and bending moment diagrams for the frame.

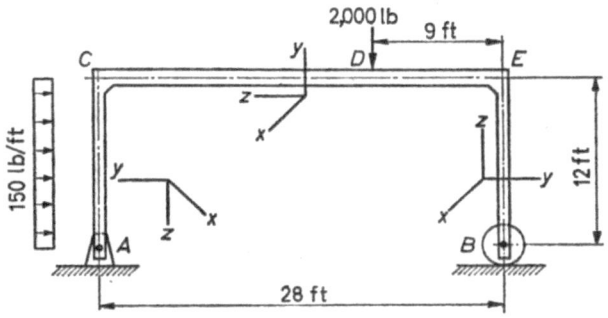

Figure 5.24. Frame of Problem 5.1

ANS:

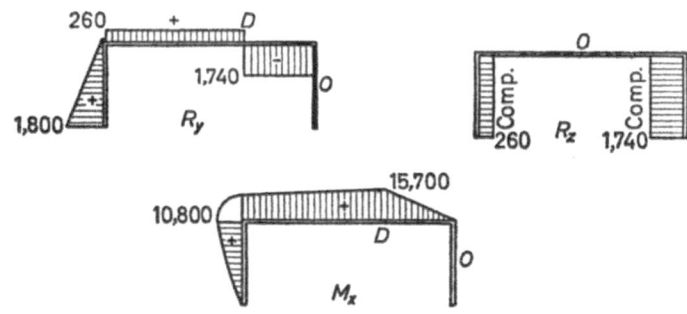

Figure A5.B1. Answer to Problem 5.1

Problem 5.2

In Problem 4.2, *Figure 4.36*, determine bending moments through-out structure.

ANS:

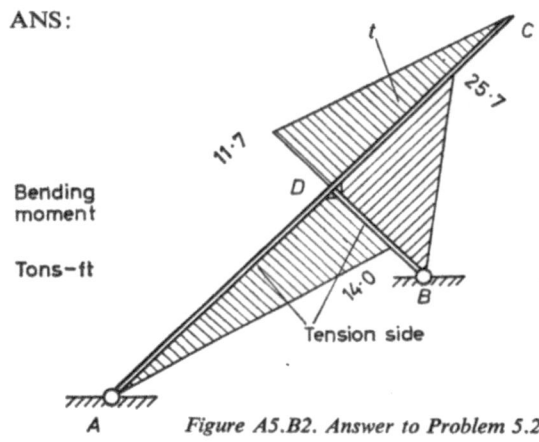

Bending
moment

Tons-ft

Figure A5.B2. Answer to Problem 5.2

Problem 5.3

In *Figure 5.25* a ramp structure is shown in outline. What are maxima values of shearing force, axial thrust and bending moment and where do these occur?

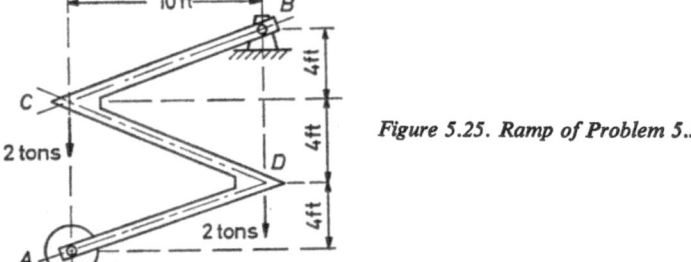

Figure 5.25. Ramp of Problem 5.3

ANS: Shearing force 1·86 tons, *A* to *D* and *C* to *B*
Axial thrust   0·75 tons compression *A* to *D*
              0·75 tons tension   *C* to *B*
Bending moment 20 tons-ft   *C* to *D*
(Shear and thrust are zero on length *CD*.)

Problem 5.4

In truss of Problem 4.7, *Figure 4.41(a)*, find forces in all members.

ANS: *AC* 3·60 tons compression
AD 4·24 tons compression
CD 5·05 tons tension
CB 2·14 tons compression
DB 1·23 tons compression

Problem 5.5

Plot the bending moment diagram for the Gerber beam of Problem 4.5, *Figure 4.39*. State location and magnitude of the maximum bending moment.

ANS:

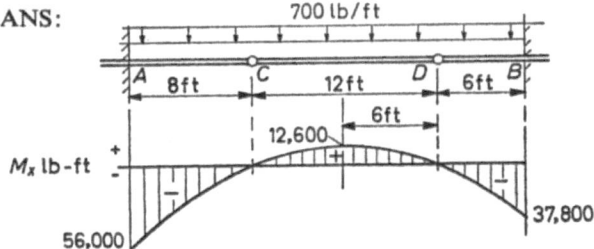

Figure A5.B3. Answer to Problem 5.5

ANS: At *A*; −56,000 lb-ft

Problem 5.6

A church roof is in the form shown in *Figure 5.26*. For this three-hinged frame, draw out the axial thrust diagram.

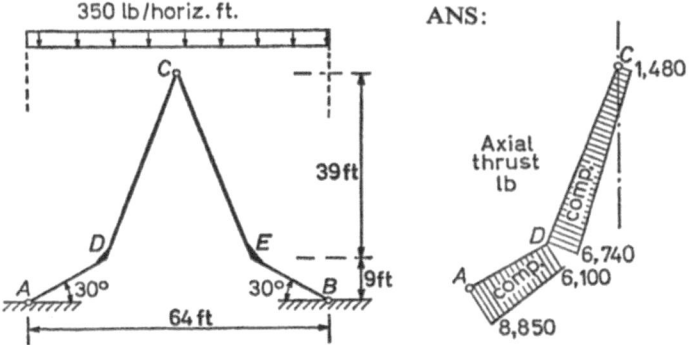

Figure 5.26. Three-hinged frame of Problem 5.6

Problem 5.7

Find forces in all members of structure of Problem 4.6, *Figure 4.40*.

ANS: *AC* 3·12 tons compression
AD 0·80 ton tension
DC 2·16 tons tension
CE 2·16 tons compression
CB 1·18 tons tension
BE 0·80 ton compression

Problem 5.8

In Chapter 4, Example 4.15 deals with a four-hinged frame, stiffened by the insertion of two corner ties as shown in *Figure 4.34*. Draw out shearing force, axial thrust and bending moment diagrams for leg *AC*.

ANS:

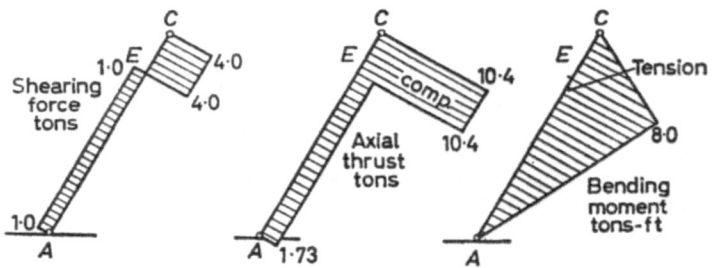

Figure A5.B4. Answer to Problem 5.8

Problem 5.9

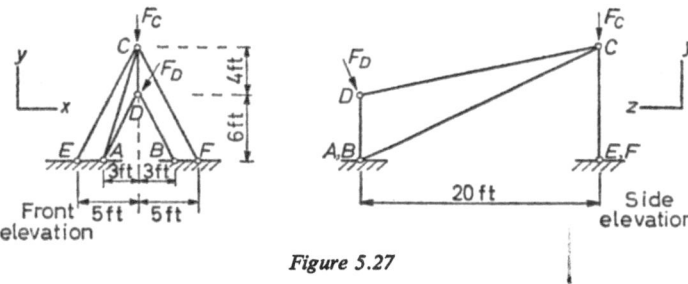

Figure 5.27

In space truss of *Figure 5.27* find forces in all members. F_C is vertical, 3 tons. $F_{Dx} = -1$, $F_{Dy} = -4$, $F_{Dz} = -2$ tons.

ANS: DA 3·56 tons compression
DB 1·33 tons compression
DC 2·04 tons compression
CA 2·25 tons tension
CE 2·32 tons compression
CF 1·68 tons compression

Problem 5.10

Refer back to Problem 3.3, *Figure 3.12* and determine the location and value of each of the maxima of shearing force, axial thrust and bending moment in the structure.

ANS: Shearing force   1·79 tons on BG
Axial thrust   1·79 tons compression on BD
or 1·05 tons tension on AC
Bending moment 6·98 tons-ft at E

Problem 5.11

In Problem 3.6, *Figure 3.15*, describe variations of M_x, M_y and M_z along *BG*.

ANS:

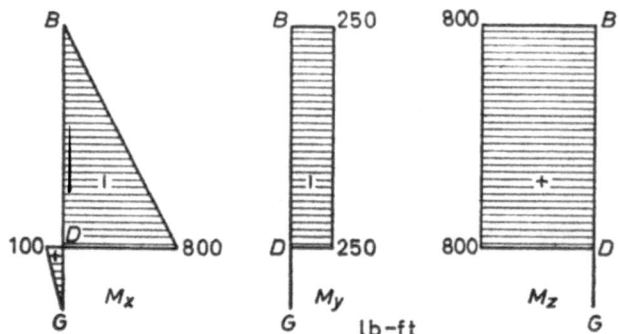

Figure A5.B5. Answer to Problem 5.11

Problem 5.12

Problem 3.7, *Figure 3.16* shows a pitched beam with a triangular distributed load. How much is the maximum axial thrust and the maximum bending moment?

ANS: Axial thrust 2·42 tons compression from *A* to *B*
Bending moment +12·92 tons-ft, at mid-span

INDEX

Acceleration,
  due to resultant, 2
  linear, 4
  rotational, 4
Articulation,
  frames, 122
  full (trusses), 111
  link supports, 101

Balancing force, 31, 59, 70
Bearing pressure, 20
Bending moment, 12, 56
Bow's notation, 174

Centroid, -line, 135
Co-ordinate system,
  choice of, 58, 136
  left-hand screw, 59
  right-hand screw, 58

Determinacy, statical, 96
Diagrams, force,
  axial thrust, 141
  bending moment, 141
  rules, 142, 150
  shearing force, 141
  signs, 148
  torsional moment, 147

Equilibrium of pin,
  method of, 166
Equilibrium of systems, 6

Fluid mechanics, 1
Force(s),
  auxiliary system of, 35
  balancing, 31, 59, 70
  diagram of, 33
  drag, 60
  external, 6
  internal, 9, 137, 140
  lift, 60
  manipulation of, 32
  reaction, 7
  shearing, 12, 56
Frames,
  three-hinged, 103
  four-hinged, 124
Freedom,
  degrees of, 97

Gerber beam, 107

Hinge,
  force, 102
  inserted, 103
  plane and sliding, 110

Indeterminacy, statical, 96

Just stiff truss,
  plane, 112
  space, 118

Link polygon, 38

189

Loads,
  concentrated, 16, 140
  dead and live, 6
  distributed, 17, 148
  dynamic, 7
  equivalent distributed, 17
  miscellaneous, 153
  vertical only, 101

Mechanics, branches, 2
Moment,
  pitching, 62
  rolling, 62
  yawing, 62

Newton's laws, 2, 5

Plane equilibrium, 60
Pressure,
  centre of, 60, 78
  definition, 18
  diagram, 22
  resultant—force, 21
Principal axes, 135

Reactions (see Supports)
Redundancy, 99
Resultant
  as couple, 3
  of,
    force system, 3
    two co-planar forces, 33
    two space forces, 35
    three co-planar forces, 37
    three space forces, 44
  pressure force, 21
Rigid bodies, 2

Scalars, definition of, 30
Sections, method of, 166

Stable bodies, 96
Strength,
  bending, 12
  compressive, 13
  of material, 12
  shearing, 12
  tensile, 13
Stress,
  axial, 24
  bending, 24
  diagram, 118, 174
  in pressure vessel, 19
  shearing, 24
  torsional, 24
Support(s),
  in,
    plane, 84
    space, 77
  reactions, 73
  restraints, 97
  types of, 72, 77

Transfer of loads, 11
Transposition,
  application of method, 47
  principle, 35
Trusses,
  internal forces in, 166
  graphical solution of, 174
  plane, 112
  space, 118
Twisting moment (torque), 56

Unstable bodies, 96

Vector(s),
  addition, 33
  definition of, 30
  free and fixed, 30
  manipulation of, 32
  subtraction, 34